Ben Franklin Stilled the Waves

Charles Tanford is James B. Duke Professor Emeritus of Physiology at Duke University. He is the author of *Physical Chemistry of Macromolecules, The Hydrophobic Effect: Formation of Micelles and Biological Membranes*, and coauthor with Jacqueline Reynolds of *Nature's Robots: A History of Proteins*, which have become classic works in their field.

He is the author of more than 200 research contributions, is a member of the National Academy of Sciences, and in 1979–80 was president of the Biophysical Society. Professor Tanford now lives and writes in England.

Ben Franklin Stilled the Waves

AN INFORMAL HISTORY OF POURING OIL ON WATER WITH REFLECTIONS ON THE UPS AND DOWNS OF SCIENTIFIC LIFE IN GENERAL

CHARLES TANFORD

OXFORD
UNIVERSITY PRESS

Great Clarendon Street, Oxford OX2 6DP

Oxford University Press is a department of the University of Oxford.
It furthers the University's objective of excellence in research, scholarship, and education by publishing worldwide in

Oxford New York

Auckland Bangkok Buenos Aires Cape Town Chennai
Dar es Salaam Delhi Hong Kong Istanbul Karachi Kolkata
Kuala Lumpur Madrid Melbourne Mexico City Mumbai Nairobi
São Paulo Shanghai Taipei Tokyo Toronto

Published in the United States
by Oxford University Press Inc., New York

First published in hardback by Duke University Press 1989

Published in paperback by Oxford University Press 2004

British Library Cataloguing in Publication Data
Data applied for

ISBN 0 19 280494 4

1 3 5 7 9 10 8 6 4 2

Typeset by RefineCatch Limited, Bungay, Suffolk
Printed by Clays Ltd, St Ives plc

Foreword

My curiosity was piqued fourteen years ago when I saw Charles Tanford's *Ben Franklin Stilled the Waves* on a shelf of new books. I had known Tanford as a distinguished biophysicist and physical chemist and had learned much from his earlier scientific works. What did Tanford have in mind in writing about Franklin and the pouring of oil on water? The unconventional title prompted me to read the book, which proved to be enlightening and delightful. A few months ago, I was gratified to learn that a paperback edition was planned and was pleased to be invited to comment on a book I highly value.

The cells of living organisms are bounded by a fluid membrane that gives them their individuality by separating them from their environment. The core of all biological membranes is a closed sheet just two molecules thick, called a lipid bilayer, which serves as a barrier to the flow of matter between the inside and outside. Proteins embedded in the lipid bilayer mediate distinctive functions by serving as pumps, channels, receptors, and energy convertors. How was the lipid bilayer, a fundamental motif of life, discovered? Tanford provides a charming, informative, and highly accessible account in which Ben Franklin occupies center stage.

Tanford begins with Franklin's letter of 1773 to a scientific friend in which he mentions Pliny's first-century AD description of the practice of seamen of his time to still the waves in a storm by pouring oil into the sea. Franklin then tells of his own wave-stilling experiences before describing an experiment he performed at an English country pond, in which he observed a "sudden, wide, and forcible spreading of a drop of oil on the face of the water". Tanford is a superb guide and storyteller in leading us from Franklin's remarkably simple and trenchant experiment to the establishment two centuries later of the lipid bilayer nature of membranes.

In following the trail, we are rewarded with vignettes of scientists who made key discoveries and vivid descriptions of the scientific and social context of different eras. I especially enjoyed the sketches of Lord Rayleigh and Irving Langmuir, which highlight their incisive and enduring contributions to our understanding of surfaces and contrast their very different scientific and personal temperaments. The account of Agnes Pockels' key experiment, carried out in her kitchen, and of her correspondence with Lord Rayleigh, is sensitive and moving. Tanford also enriches and informs by highlighting connections made and connections missed between different fields of science.

Reading *Ben Franklin Stilled the Waves* a second time was much like savoring wine that has become richer and deeper with the passage of time.

LUBERT STRYER

Stanford University
February 2003

Preface to the Paperback Edition

This book was originally written about fifteen years ago, some time after it had been established that a phospholipid bilayer is the dominant structural component of a cell membrane and that membrane proteins represent a special category, preferentially associated with such a bilayer and usually not soluble in water or simple aqueous solutions. Moreover, it had become clear that the ultimate molecular basis for all this, both the propensity of phospholipids to exist in the bilayer form and the vicinal preference of membrane proteins, must be the hydrophobic/hydrophilic dichotomy—molecular orientation arising from the demanding physical forces between water molecules and their neighbours in the biological milieu. The very existence of the living cell and the segregation and partitioning of its contents can basically be viewed as dictated by this factor, providing a physico-chemical link between the living cell and the many phenomena of the inanimate world, in which the hydrophobic force plays a similar determining role, without, of course, any lipid or protein being involved.

Who would have guessed that there might be a parallel between one of the most basic features of "life" and such a simple phenomenon as the spreading of oil on water and the stilling of the waves that results from it? What is the history of research and comprehension into this and closely related questions? In tracing the answer, we find that a fascinating assembly of people have, each in their own way, independently taken the lead along a similar trail: Nobel laureates Irving Langmuir and Lord Rayleigh, the amateur Agnes Pockels, who worked in her kitchen, the Dutch pediatrician Evert Gorter, and (perhaps most fascinating of all) the American statesman and polymath Benjamin Franklin. Physics, chemistry and biology have all been enriched by their work.

This paperback reissue is essentially identical with the original

1989 edition published by Duke University Press. A few small corrections have been made to eliminate some errors that were in the original; in other respects the text remains unaltered. Two new items have been added to the wide-ranging bibliography—because they should have been there all along—but no attempt has been made to bring the list up-to-date by including more recent work.

C. T.

Easingwold
February 2003

Acknowledgments

My vague idea of some years ago, to write "something" about Ben Franklin's experiment and its relevance for biology, would never have come to fruition without the constant urging and encouragement of Jacqueline Reynolds. She also contributed to the necessary research and to the endless quest for little errors of fact in the text. I am deeply grateful to her. I also thank Ben Reynolds for his invaluable advice, especially on style; Seymour Mauskopf for setting me straight on central themes in the history of chemistry; J. J. Hermans for his recollections of Evert Gorter; John Edsall for his recollections of Lawrence Henderson; George Gaines and the General Electric Company for the photograph of Irving Langmuir; and the Rijksuniversiteit te Leiden and Drs. J. G. Leroy and J. Kint of the Rijksuniversiteit Gent for different versions of the photograph of Evert Gorter.

I am grateful to Cambridge University Press for permission to reproduce a portion of Clarence Greig's delightful translation of Pliny the Younger's letter about his uncle's death at Pompeii. I thank Oxford University Press for permission to reproduce figure 22.

Contents

CHAPTER ONE

Introduction

> In these experiments, one circumstance struck me with particular surprize. This was the sudden, wide, and forcible spreading of a drop of oil on the face of the water, which I do not know that any body has hitherto considered. If a drop of oil is put on a polished marble table, or on a looking-glass that lies horizontally, the drop remains in place, spreading very little. But when put on water it spreads instantly many feet around, becoming so thin as to produce the prismatic colors, for a considerable space, and beyond them so much thinner as to be invisible, except in its effect of smoothing the waves.
>
> Benjamin Franklin, letter to William Brownrigg, November 7, 1773

1. Pouring Oil on Troubled Waters

THE Venerable Bede tells the story of a worthy priest commissioned to escort the Princess Eanflaed from Kent to Northumberland to become the bride of King Oswiu. The journey was to be by sea, and the priest, concerned about his young charge's safety and comfort, sought the blessing of his superior, the Bishop Aidan, which he received. Aidan also gave him some oil to use in case of storm. "Remember to pour the oil I have given you on to the sea; the winds will drop at once, the sea will become calm and serene, and will bring you home the way you wish."

This is, of course, not the first mention of the effect of oil on the waves of the sea. The phenomenon must have been observed by the earliest seafarers who used oil to cook meals on board ship and who would have been familiar with the darkened smooth surface, devoid of whitecaps, that appeared in the ship's wake when greasy residues were dumped overboard. Ben Franklin himself first read about oil on water as a youth in Pliny's *Natural History*, the great encyclopedia of fact and fiction originally published in the first century A.D. What is important to us here is that Franklin saw something more than the calming of the waves when he first verified the observation experimentally on the pond at Clapham Common around 1770, something that all his predecessors must also have "seen" but not remarked, whereas Franklin was struck by it "with particular surprize." This "something"—the *spreading* of the oil on the water to a very large area—is the reason for this book.

Franklin sensed the importance of the spreading of the oil, that it could reveal crucial facts about molecules and the forces between them, if only the questions that the observation raised could be answered. He was able to perceive these questions only dimly, for the time was not quite ripe to do more, but subsequent scientists have done exactly the same experiment (altered only to reduce it from pond to laboratory scale), and each one in turn has learned more from it. Rayleigh, the physicist, measured molecular size. Langmuir, dean of surface chemists, learned about molecular shape and contortions. And then came the turn of biologists, who recognized its relevance to the core of biology, the very existence of the living cell.

We are often told that science progresses by the development of new and ever more sophisticated methods. Here we have something different: the same simple experiment, repeated again and again, and we learn more from it as our vision and imagination are broadened. And because the experiment is so simple, we need no special training for our own insight to deepen as we go from chapter to chapter, from 1770 to the present day.

2. The Reverend Mr. Farish

Coincidence, chance encounters, "Fate." These are common themes in mythology and literature, but they rarely intrude into scientific history. Yet science, too, is a human endeavor, subject at times to the vagaries of blind Fortune—and so it happens here. Were it not for the Reverend Mr. Farish's fussy curiosity, Ben Franklin's experiment would not have been recorded in the scientific archives.

The Reverend Mr. Farish, not otherwise identified in the original paper or in more recently published commentary, is mentioned many times in the Call Book of the Carlisle Diocese, now preserved in the Cumberland County archives. He was James Farish, vicar of the Church of St. Michael the Archangel, in Stanwix, Carlisle, a Saxon church at the time, with an ancient square tower. He was clearly a man of some stature, sometimes asked to preach in the Cathedral Church in Carlisle when all the ministers of the diocese met there in assembly. Like many churchmen of his day, Mr. Farish took a keen interest in science. He made no original experimental contributions, as Joseph Priestley did and other pastors we shall meet later in this book, but he absorbed with naive enthusiasm all the latest scientific advances and undoubtedly shared them with his parishioners.

Farish's credulity balked when he first learned by word of mouth about Ben Franklin's demonstration of the oil-spreading experiment for Dr. William Brownrigg on nearby Derwent Water lake. He found it "a little incredible" and, on hearing accounts from other eyewitnesses, suspected them all "of a little exaggeration." Nor can we blame him; even today Franklin's observation strikes us at first as rather incredible.

Mr. Farish was sufficiently concerned to write a letter to Dr. Brownrigg, requesting an authentic account of the experiment. Brownrigg forwarded the letter to Ben Franklin ("By the enclosed from an old friend"), and Franklin responded with a detailed report of his recollection of the experiment ("Perhaps you may not dislike to have an account of all I have heard, and learnt, and done . . ."). Brownrigg had Franklin's letter published in the *Philosophical Transactions* of the Royal Society—England's prestigious scientific academy—in 1774, which is why we know about the experiment today and about Franklin's ideas concerning his observation.

As we shall see, Mr. Farish's letter actually shows him to be a rather foolish gentleman, who did not learn a lesson he might have learned from Franklin's experiment. He was an unimportant figure, worthy of historic mention only for this one accident, that his letter was directly responsible for the publication of Franklin's observations.

But there is a funny thing about this. Our thoughts sometimes return to foolish Mr. Farish in the course of this book. For scientists, too, can be guilty of mindless folly. And, surprisingly, sophisticated and well-trained professionals can be subject to the same kind of mental lapse that afflicted poor Mr. Farish, the failure to learn a lesson that was written in plain language for all to read.

3. Lords and Ladies and Even a Pediatrician

Why must science always be represented as impersonal, devoid of human interest? We can't imagine art without the artist, are taught in fact that art is an expression of the artist's innermost self. Yet science is supposed to be different, purely factual, divorced from the scientist's private life. Popular science fiction shows on television (like "Star Trek") have mechanical robots with a more abundant measure of human warmth than is commonly ascribed to human scientists, past or present. Is this one of the reasons why the general public prefers science fiction to science fact? I can't answer that, but I can speak for the fact that people are important to the scientific story of oil on water—not only the Reverend James Farish, the accidental instrument of there being a story to tell, but the scientists themselves. They have human virtues and failings, like anyone else, and they embellish our history thereby.

The eighteenth century was the Age of Enlightenment, which is not to be interpreted passively, as *being* enlightened, but rather actively, as an advocacy of *becoming* ever more enlightened, especially about the natural world around us. There were few professional scientists, almost none in England, but it was fashionable to be an amateur scientist, to report on nature as seen on the local moors, to extol the virtues of Newton, to form discussion groups that listened avidly to travelers' tales of distant lands. Anyone who chose to exercise his brain and powers of observation could learn something new about almost any subject, and countless important contributions to science were in fact made by amateurs without special training. For example, the celebrated Swedish taxonomist Linnaeus, one of the few professionals of the time, firmly believed that swallows hibernate *under water* when they disappear from sight in the fall. It was the English amateur Peter Collinson (close friend of Ben Franklin's) who proved otherwise.

Benjamin Franklin is in a sense typical of the age: a self-taught scientist, inventor, and writer, a homespun philosopher, and at the same time a dedicated statesman and negotiator. He stands out, of course, as one of the great men of all time in the English-speaking world, rarely matched in intellectual power and clarity of expression. But many of his friends and associates engaged in equally varied pursuits and were equally prolific in correspondence and debate. We shall meet some of them a little later in the book and shall take delight in their adventurous spirit, their lack of timidity in word or thought.

What is perhaps surprising as we follow the story of oil spread on water is that those who came after Franklin, one hundred years later, even 150 years later, turn out to be equally fascinating people. Science as a whole had become increasingly specialized, increasingly dependent on complex apparatus. Scientists were becoming important figures, making prophecies as they stood before

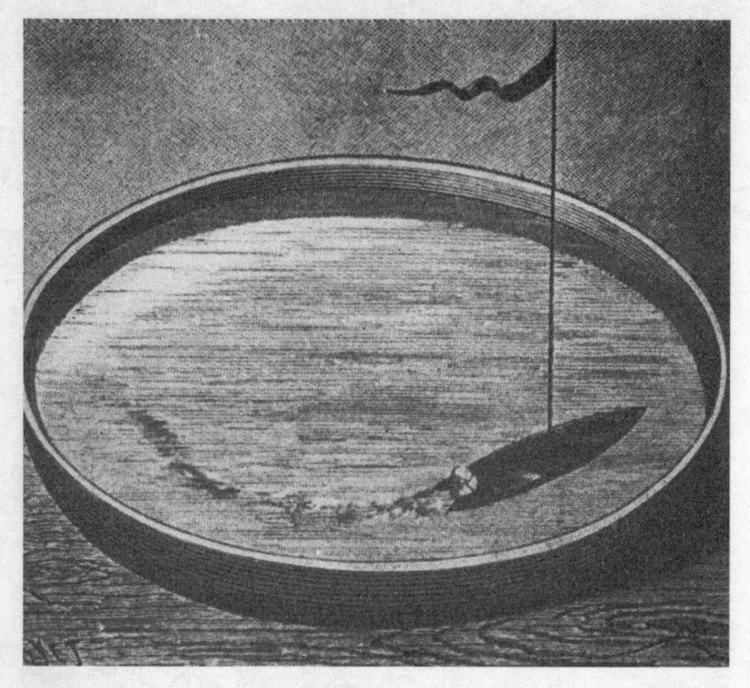

Figure 1. One of several methods that can be used to detect the spreading of a drop of oil is by means of tiny chips of camphor. They move about rapidly on a water surface, but lie still on oil. The French scientist Henri Devaux attached his camphor chips to a little tin boat, complete with mast and colored banner. This photograph is from a 1932 review article by Devaux in *Kolloid Zeitschrift*.

machines with dials and gauges and flashing lights. I suppose that it required a somewhat idiosyncratic character to resist this surge towards instrumentation and to invite one's colleagues' ridicule by spending one's time sailing toy boats in a tub full of water (Fig. 1).

These latter-day practitioners of surface chemistry reflect a lingering of the adventurous spirit of the Enlightenment—discarded by much of the scientific world in favor of more purposeful, predictable investigation. They were a motley crowd: Lord Rayleigh, an altruistic English lord, who built a laboratory in his home and financed all his research out of his own pocket; Agnes Pockels, a modest gentle lady from Braunschweig, one of the earliest women to make lasting contributions to science (preceding Marie Curie by a few years), whose research was done in her own kitchen; Evert Gorter, a renowned Dutch pediatrician; Irving Langmuir, one of America's greatest industrial scientists. Their nationalities, social backgrounds, primary occupations, and income levels were different. They never met and talked to each other as a group. But they had in common the conviction that one could still learn something by measuring the area to which a drop of oil expands on water. And they were correct. The group includes three winners of the Nobel Prize: Rayleigh, Langmuir, and W. C. Roentgen, who made some forays into similar research before he accidentally discovered X rays.

There are fools and villains too, as in any tale of human endeavor, but few of them are memorable. Justus von Liebig, the famous chemist, appears briefly as a malevolent despot with little tolerance for ideas other than his own. Lawrence Henderson, respected early in this century as a "natural philosopher in the broad sense," appears less admirable in retrospect, more of a theologian than a serious scientist. There are some others, but they are less significant. One is, however, led to wonder how any of them could have attracted a devoted following, as some of them undoubtedly did. And why did it take so long for some of the "good guys" to be appreciated?

4. The World Outside

It is difficult today to imagine how it was in Franklin's time, around 1770, without electricity, before steamships, before telegraphy. They were in fact stirring years, as dramatic as any in the modern history of the Western world.

Central to the life of Ben Franklin, of course, was the increasing alienation between England and its American Colonies. That is what brought Franklin to England as an emissary. He was an Anglophile as well as an American patriot, an admirer of English institutions, genuinely hopeful that he could persuade England to redress legitimate grievances, never doubting that the Union could be preserved. He was about to be sorely disillusioned, in a humiliating and vindictive hearing by the Privy Council. In the meantime there was the tedium of waiting for an answer to his petitions. Franklin used the time for extensive travel, he began his never-to-be-finished autobiography—and, one windy day, he poured some oil on the pond at Clapham Common. The story of Franklin's mission has been often told, but it bears repeating, and I shall summarize it again in the following chapter.

London, strangely, seemed relatively uninterested in the disputes with the Colonies. James Boswell was going on his social rounds, describing the London scene in his delightful journals. He speaks of Franklin several times, but not of his political mission. Shakespeare's plays, though nearly two hundred years old, still occupied center stage in London. David Garrick, indisputably England's greatest actor of the century, was their most famous interpreter and Johnson's and Boswell's long-time friend. Joshua Reynolds was another of Boswell's friends, a painter of portraits of the socially prominent, and the first president of the newly founded Royal Academy of Arts. In 1773 Boswell left London altogether for a time and went with Samuel Johnson on their celebrated journey to the Hebrides. Perhaps surprisingly, *everyone* was eager to travel (including Ben Franklin)—despite poor inns and uncomfortable means of transportation.

Ironically, though England was unimpressed by the American struggle for liberty, it was greatly concerned about a war of independence on the little island of Corsica, seeking freedom from domination by the city-state of Genoa. Boswell wrote a book about this war, and London's literary society lionized the revolutionary leader Pasquale Paoli when he sought refuge in London after the revolution failed. In retrospect, concern about Corsica may have been justified by subsequent events. Corsica was annexed by France, its inhabitants became subject to service in the French army, and one Corsican who would shortly become a corporal in France by that route was named Napoleon Bonaparte.

In France itself, there was splendor at the court and much more official encouragement of science than in England. But revolution was brewing there too, uglier than in Corsica. France's most famous scientist, Antoine Lavoisier (a personal friend of Ben Franklin's), would lose his head to the guillotine in 1794.

Farther east in Europe, the Enlightenment came more slowly than in the West. Catherine the Great, after an aborted attempt at liberalizing legislation, reverted to a despotic reign as Russian empress. But even in Russia Ben Franklin's name was familiar. A Swedish physicist, Georg Wilhelm Richman, was killed by a severe lightning bolt in St. Petersburg in 1753 when he attempted to repeat Franklin's demonstration of the marvelous identity of lightning and electricity to the Russian Academy of Science.

Maria Theresa reigned, at least in spirit, over Austria and made Vienna a center for art, music, and elegance. Haydn was court musician for the Counts of Esterházy from 1761 to 1790. Mozart was fourteen in 1770 and already becoming well known; in Salzburg you may visit the house where he was born and the larger quarters to which his family moved in 1773—about the time of Franklin's letter.

It is not known if Ben Franklin ever heard of Mozart, but Mozart may well have known Franklin by name. Franklin's genius for invention penetrated to the world of music. He refined the German "Glockenspiel" (made from beer glasses) into a more delicate instrument which he called an "armonica," and which became quite fashionable in central Europe for a while. Several famous composers wrote music for the instrument, including Mozart, who composed a famous quintet for armonica, flute, oboe, viola, and cello in 1791.

The world around us changes as time advances, and we shall try to catch a glimpse as we pass through the years: of London in 1890 and of America around 1910. The contrast with Franklin's London and America is spectacular, and the contrast between science then and now is equally spectacular. In the laboratory we have fancy instruments now, to measure practically anything we could ever want to measure and much faster than ever before. Furthermore, the number of scientists and their publications have increased exponentially.

Strangely, the pace of scientific progress does not seem to have accelerated as much as one might have expected. The human mind seems to be the rate-limiting factor more than technology and the ease of data acquisition.

5. Connecting Threads

All those who wander are not lost. Trails can branch off in unlikely directions, inviting exploration, leading to unexpected vistas. So also with the intellect. Connecting threads join seemingly unrelated facts or ideas, revealing (if followed) new insights, creating new patterns in the mind. Discovery of unity in diversity—that is perhaps the secret of creative genius.

To put it another way, we might make the distinction between pure chronicle on the one hand and scholarly synthesis on the other. A chronicle recounts events in the order in which they happen. It may well require extensive research, the delving into old records and private correspondence to find out exactly what did happen. A chronicle may well be exciting as it unfolds before us in the pages of a book or on our television screen. But synthesis is more than a chronicle. Connecting threads create links across gaps of time and space, and links within the mind among different philosophies and subject matters. What emerges is far more than the sum of its parts, like a symphony that could never be imagined from the individual parts alone.

Connections will in fact be a keystone of this book. Some have already been alluded to in this chapter—the contention, for example, that what an individual does scientifically is related to who he is and to the world in which he lives. We cannot now spell out the exact rules for such a relation; "synthesis" is too grand a term for these connections at the present time. Nevertheless, they entertain and stimulate, give us food for thought, provide the basis perhaps for some future, not yet conceivable, insight.

A more traditional link has been made by going back in time from Franklin's experiment to early mariners and forward to modern pourers of oil on water. In this time domain, the spreading of oil on water affords a marvelous opportunity to focus on age-old problems in a broad sweep of science, questions which each succeeding generation, it seems, must grapple with anew. For example, do atoms and molecules really exist? The answer has wavered back and forth from the time of the Greeks almost up to the present. The period of Franklin's experiment represented a kind of window in time, when "ultimate particles," with empty space between them, were pretty much taken for granted. (And that is probably one reason why Franklin's paper is comfortably readable by a modern audience.) But the dispute arose again late in the nineteenth century and was not truly resolved until the basic particles of matter—or their tracks through space—were actually made individually visible.

Another eternal question is what constitutes the molecular basis for life. At first it seemed that *organic chemistry* held the key. In the last few decades the answer has been sought in the field of *genetics*, in the genetic material (DNA) that carries the information for creating a living organism and in the means whereby this information is processed and transmitted. In this book (and intimately related to the spreading of oil on water) our attention is focused on yet another part of the problem, on *containment*. How do we form a boundary between life (one specific individual living thing) and non-life? How do we cross that boundary to carry in food or signals of danger or joy?

The time domain also brings us face to face with *failed connections*, both narrowly, as revealed in the work of a single individual, and more broadly, as a kind of communal blindness. The poet Robert Frost alludes to this when he writes that Nature is always hinting at us, over and over again, but we don't notice; our minds somehow not trained to pay attention.

> How many times it thundered before Franklin took the hint!
> How many apples fell on Newton's head before he took the hint!

There are several examples in this book of "hints" that were never taken, without obvious explanations. In chapter 8, for example, why did Franklin fail to calculate a numerical value for the thickness of the oil film he had observed, despite his evident excitement over how very very thin it must be? And in the modern era (chapter 18), why was comprehension of the physical basis for cell membrane organization so painfully slow?

The very latest research in particle physics has given us a vivid example of the positive value of connections to earlier times (in this case, back to the very beginnings of physics) and across intellectual space to different problems. The research concerns *quarks*, the ultimate particles of matter of the 1970s and 1980s, the building blocks for protons and neutrons and (ultimately) atomic nuclei. Quarks can't be seen, of course; they can only be *imagined* to exist on the basis of atom-smashing experiments in the giant demolition machines we call particle accelerators. Just like atoms, back in the time of Democritus! They, too, could not be seen and had to be imagined. The analogy may at first seem inappropriate, for quarks are part of the relativistic world, where comfortable Newtonian concepts like mass and momentum have become blurred. Nonetheless, some of the same questions arise. Are quarks real? Are they themselves divisible? Here is a "popular" description of an important step in answering some of those questions, from a book by Richard March, *Physics for Poets*: "At this point, Richard Feynman intervened in a decisive fashion. Though he had not taken part in shaping the original quark theory, he reminded his colleagues that if it was to serve as the basis for a quantum field theory of matter, the real test of atomicity should be taken *not from Democritus but from Boscovich*; the crucial question was. . . ."

To quote the actual question here would take us too far afield. What is noteworthy is that Democritus lived in the fifth century B.C., and that Boscovich was an eighteenth-century contemporary of Franklin's. We shall refer to both of them in chapter 6, for the same enigma, Democritus or Boscovich, puzzled Franklin's friend, Joseph Priestley, and might have had an influence on Franklin himself.

Connecting threads, I hope, will make this book a little more than just the story of oil on water: a commentary, to some small degree, about all of science and its place in our history.

CHAPTER TWO

Benjamin Franklin

1. Gospel of St. Benjamin

WHAT is truly astonishing, most difficult to relate to today's world, is that such a giant figure in the world of men and nations as Benjamin Franklin should ever have engaged in simple, prosaic scientific experiments at all; and that he should have done so and described his results in such fashion that we can still be genuinely fascinated by them today.

I am not speaking here of the famous kite experiment, because that was a grand spectacle, a daring (and dangerous) effort to solve a great mystery of heaven and earth. It promised acclaim from scholars and from the public alike. It was in character for a man of destiny.

My reference is not to that, but to the many more humble experiments he did, which can have had no other motive than the satisfaction of an intense curiosity. The wave-stilling experiment is one of these, but there are others, such as his measurement of the temperature of the Gulf Stream during his transatlantic voyages. These experiments are always carefully thought out, and details of experimental procedure are always carefully described. (And, usually, they were not quite right the first time, and he tells us why and what minor changes he had to make.) They truly engender an instant bond of recognition with anyone who has ever done humble little experiments himself. There's somebody we would have liked to have in our own laboratory!

Numerous biographies of the "great" Benjamin Franklin exist

including a short *Autobiography*, which covers only the early part of his life. There are also many volumes of comment, not all complimentary. The title of this section, "Gospel of St. Benjamin," is taken from a modern essay by Daniel Aaron on the *character* of the man. The bibliography at the end of this book gives detailed reference to this and to a selection of other biographical works.

2. The First Fifty Years

Benjamin Franklin was born in the city of Boston in January 1706, the fifteenth of seventeen children of Josiah Franklin, ten of whom were born to his second wife, Abiah Folger. Josiah had emigrated to America in 1683 and was a tallow chandler and soap boiler by trade, with a shop in Milk Street. Benjamin received only minimal formal schooling and was assisting his father in his shop at age ten. It was evident to his father that Benjamin had a literary bent and was fond of books, and he therefore formally indentured him at age twelve to his older brother James, who was in the printing trade. There Benjamin used every penny he could lay his hands on to buy books. He cultivated friendships with booksellers' apprentices to obtain the loan of books. He chose to forego meals at his brother's house to have more time to read books, and he evaded church when he could to be alone in the printing shop. He devised ingenious trial-and-error methods to develop a good writing style. Ben Franklin became a well-read youth at a time when literacy was rare.

An indenture contract was normally binding on a young apprentice until he reached age twenty-one, but Benjamin managed to escape after a few years. His brother saw to it that he could not thereafter obtain employment in the printing trade in Boston, but Benjamin settled in Philadelphia where he rapidly became quite well-known (because of his literacy and possession of a few books) as a clever youngster, and where he was even befriended by the Governor of Pennsylvania, Sir William Keith.

Figure 2. Benjamin Franklin at age sixty-seven, painted by David Martin in London. The painting is now in the possession of the Pennsylvania. Academy of the Fine Arts.

It is worth noting at this point that Pennsylvania's colonial status was different from that of Massachusetts, which later proved to be an important factor in the shaping of Ben Franklin's career as patriot and statesman. The earliest settlers in the colony had been Swedish and Dutch, but in 1681 King Charles II gave the entire region to William Penn in payment of a 16,000 pound debt owed to William Penn's father. A Quaker by religion, Penn used the colony to provide a haven of tolerance for his fellow Quakers, and by the time Franklin moved there most of the settlers were of English extraction, as they were in Massachusetts. But the legal distinction, that Pennsylvania was private property of an absentee landlord (William Penn's heirs), would later on raise issues unique to that colony.

Governor Keith suggested to Franklin that he should have his father in Boston provide the capital that would be needed to set up a printing shop in Philadelphia, promising to give him the trade that would keep him in business. Franklin's father was not inclined to part with his money on such a risky venture, and Keith himself then magnanimously offered to help. In 1724 (at age eighteen) Benjamin Franklin set off for London on the annual Philadelphia-London boat, to buy the necessary equipment (as his brother James had done many years earlier), buttressed by the promise of a subsidy from Governor Keith and letters of introduction to people who could advise him. As it turned out, the trip would most likely have meant disaster or oblivion for a lesser man, for the Governor's promises turned out to be a cruel hoax, and Franklin arrived in London with neither money nor letters. The Governor was not really malicious in this, or at least Franklin did not think so when he wrote his *Autobiography* many years later. "He wished to please everybody," he wrote, "and, having little to give, he gave expectations."

Franklin appears not to have been unduly set back by the evaporation of the Governor's promises; the ability to cope serenely with perfidy against himself remained one of Franklin's traits throughout his life. Once in London, he quickly found work in a printing house and earned enough to be able to purchase his own printing outfit to take home. He also made the acquaintance of a number of notable people, including Dr. Henry Pemberton, physician and scholar, who was supervising publication of the third edition of Newton's *Principia*. Pemberton promised to introduce Franklin to Newton, but it never came about. Newton died in 1727, a year after Franklin returned to America from this first short London trip.

Franklin returned to Philadelphia a skilled printer. At first he worked as clerk in the store of a general merchant, then as journeyman for a printer, but it was not long before he acquired his own printing shop, in partnership with a friend, Hugh Meredith, whose father had the money needed to set them up in business. Franklin was a tireless worker. He found time for writing as well as printing. He founded a newspaper (*The Pennsylvania Gazette*), much of which he wrote himself, including letters under assumed names (such as "Alice Addertongue"), which he then answered as Editor. In 1730 he married Deborah Read—an undemanding but loyal and sensible young widow. He bought out his friend's interest in the printing shop the same year and dissolved their business partnership. He wrote books, and sold books, and started a library. And he began publication of the famous *Poor Richard's Almanack*, which made him rich and one of the most prominent citizens of Philadelphia. "I endeavor'd to make it both entertaining and useful," he writes in the *Autobiography*, "and it accordingly came to be in such demand, that I reap'd considerable profit from it, vending annually near ten thousand."

In 1728 Ben Franklin and some of his friends organized a club, called the Junto, which met once a week, originally at a tavern, for fellowship and intellectual stimulation. Clubs of this kind were common at the time throughout the English-speaking world, but this one remained active for a longer period than most, undoubtedly because of Franklin's stimulating leadership. Franklin had a list of ritual questions that were asked each week. Have you read any good books this week, heard any good stories? Have you seen examples of the unhappy effects of intemperance, or the happy effects of prudence, of moderation? Do you know interesting people, perhaps new-comers to Philadelphia, with whom we should seek acquaintance? The Junto members became the subscribers for the library he had started, and jointly decided on the books to be purchased. The library was moved to the State House, and its members adopted the name of "Library Company."

Later on, Franklin was to become a founder of the American Philosophical Society, a society for "promoting useful knowledge among the British plantations in America." It still exists, in an old building next to Independence Hall in Philadelphia. Franklin also promoted the creation of the Academy of Philadelphia, which in a few years evolved into the present University of Pennsylvania.

By 1743 Franklin had acquired a trustworthy business partner, David Hall, to whom he soon turned over most of the operation of his printing and publishing business. (His firm became known as "Franklin and Hall" and retained that name till 1766.) Franklin became free to make his own contributions to "useful knowledge" and to embark on a career (albeit a brief career) as inventor and scientist. His activities at this time included the invention of the Franklin stove and the electrical experiments that earned him a permanent place in the history of physics. Franklin was the first to use many now familiar electrical terms, e.g., battery, conductor, condenser, armature, positive and negative electricity. He identified the frightening force of lightning as an electrical discharge and invented the lightning arrestor, still in universal use today, which provides an early example of an immediate application of the results of "pure" basic research for a device of practical utility.

It is essential for an understanding of the free spirit of the period to emphasize that Franklin had no special training to equip him for this transformation from printer and journalist to internationally recognized scientist, and had to teach himself all that he needed to know. However, being self-taught does not mean being uneducated. Franklin had rigorously studied the science of his day, had read all the scientific writings he could obtain. He had read Newton's *Opticks*, the works of Robert Boyle, of Stephen Hales, and of experimental philosophers from the European Continent. He was helped by a stream of correspondence from Peter Collinson in London, who wrote regularly to the Library Company, to communicate everything new that had been reported at meetings of the Royal Society, of which Collinson was an active Fellow. Collinson's letters kept Franklin and his friends as up-to-date as scientific journals would do today.

Franklin's interest in electricity was probably first stimulated by Adam Spencer, one of the itinerant lecturers who went from city to city in those days, supplying as best as they could the popular thirst for knowledge. Franklin had already heard Spencer speak once in Boston, and in 1744 he invited him to lecture in Philadelphia. Shortly thereafter, Peter Collinson sent the Library Company an electrical tube as a gift, and actual experiments began as a group activity of the Company.

Franklin's first "electrical" letter to Collinson in March 1747 begins as follows:

> Your kind present of an electric tube, with directions for using it, has put several of us on making electrical experiments, in which we have observed some particular phaenomena that we look upon to be new.

Results of these and subsequent experiments and Franklin's far-reaching theoretical interpretation of them were reported in frequent letters to Collinson. Collinson read some of the letters to the Royal Society and eventually arranged for the publication in London of *Experiments and Observations on Electricity*, which was essentially a collection of the letters. This book went through several editions, was translated into French, German, and Italian, and spread Franklin's fame throughout the European scientific community. In 1754 Franklin was awarded the prestigious Copley Medal of the Royal Society for his work on electricity, a rare honor for a nonresident of Britain, and shortly thereafter Franklin was himself elected a Fellow of the Society.*

A fascinating aspect of Franklin's letters is his sense of delight and wonder: delight at being freed at a relatively early age from the tedium of business management, at being able to engage in what he clearly viewed as a less serious activity; wonder at the marvels of electricity, both as revealed by the work of others ("Musschenbroek's wonderful bottle"—later known as the Leiden jar) and by his own discoveries. There is no evidence that Franklin realized, at least initially, that he was making a solid, lasting contribution to the foundations of physics or that he could foresee the revolution that electricity would subsequently create in human society. On the contrary, in his fourth letter to Collinson (1749) he describes himself as "chagrin'd a little that we have been hitherto able to discover nothing in this way of use to mankind." He proposes to put an end to his experiments for the summer season, but with gusto—with a party on the banks of the Schuylkil:

> Spirits, at the same time, are to be fired by a spark sent from side to side through the river, without any other conductor than the water; . . . a turkey is to be killed for our dinner by the *electrical shock*, and roasted by the *electrical jack*, before a

* British citizenship was a requirement for becoming a Fellow, but Franklin of course fulfilled this requirement—the Colonies were part of Britain, and the Royal Society roster, before American independence, contains the names of several American-born Fellows. One of them was the notorious witch-hunter Cotton Mather. He is best known for his diatribes from the pulpit, but he also made many contributions to scholarly knowledge and was elected a Fellow on that basis in 1713. Another early fellow was Elihu Yale, elected in 1717. The citizenship requirement for membership still applies, not only for the Royal Society, but also for the equivalent national academies of science in other countries, including the United States. Some of them today can honor non-nationals by electing them as "foreign associates" without voting rights.

> fire kindled by the *electrified bottle*: when the healths of all the famous electricians in *England, Holland, France*, and *Germany* are to be drunk in *electrified bumpers*, under the discharge of guns from the *electrical battery*.

An *electrified bumper* is defined in a footnote as an electrified glass of wine, which gives a shock when brought to the lips "if the party be close shaved, and does not breath on the liquor." Is there a tradition somewhere that Franklin was a somber, serious individual, never given to frivolous pleasure?

Within the next year Franklin invented the lightning rod, an obvious "benefit to mankind," but there was no perceptible change in his priorities. There was not much incentive in America at the time for the pursuit of scientific studies, in contrast to France, for example, where science had an integral role in the national culture and in the activities at the court of Louis XV. In America, benefit to land and people had precedence over benefit to mankind, and Franklin soon felt, and did not resist, a strong call to civic service. He never gave up his interest in science and in his leisure hours continued to occupy himself with scientific questions, but statesmanship and negotiation became his primary activities.

By 1754 Franklin was a respected public figure. He was appointed deputy postmaster-general for all the American Colonies, and he greatly improved postal service and at the same time made it profitable for the Exchequer. He was to hold the position for more than twenty years. The salary, though small, helped keep him financially independent.

Initiation into serious diplomacy came the same year, when Franklin was appointed as one of the official Pennsylvania representatives to a congress in Albany that was convened to deal with the Iroquois Indians, to make them side with the British in the then worsening military conflict with the French. The French had, for example, captured part of the Ohio River Valley, including the unfinished British fort in present-day Pittsburgh, which they completed and named Fort Duquesne. Franklin was a conspicuous figure at the congress, intent on persuading other delegates to his own strong conviction that the several colonies needed to pool their resources if they were to keep the frontier in British hands. He buttressed his appeal with a quantitative projection of the huge growth in population that the Colonies were likely to experience in the near future, and he pointed out that the inevitable result would be a pressure for westward expansion of colonial territories. The frontier must be kept in "our" hands to prepare for that day. It is noteworthy that Thomas Malthus, the famous pioneer of the science of demography, was aware of Franklin's projection and quoted it in 1798 as part of his famous essay on population expansion.

The most serious need for diplomacy was, however, not at home, but in dealings with the Mother Country. Ben Franklin spent sixteen of the next eighteen years in London! (And, after one year at home, during which he signed the Declaration of Independence, another ten years in France.)

3. First Mission to London

British efforts in 1755 to secure the western frontier by recapturing Fort Duquesne from the French ended in disaster. Bands of unfriendly Indians raided outlying farms of German colonists, and killed their families or forced them to flee. Creation of an independent militia became a necessity, and taxes had to be imposed on the populace to meet the expenses. Taxation raises emotions at any time or place, but Pennsylvania had special problems which made the issue especially divisive.

One problem was posed by the Quakers, then as now resolute pacifists. Though a minority in the population, they constituted a majority in the Assembly. A militia bill could not be passed unless it exempted them from compulsion to bear arms, and the act that was written by Franklin (who was himself never a Quaker) and eventually adopted contained a proviso to assure this. Understandably it made some non-Quaker citizens livid with anger. Former friends became lifelong enemies: William Smith, for example, provost of the Academy of Philadelphia, became a powerful Franklin opponent for many years.

The second problem was Pennsylvania's status as a "proprietary colony," not directly under control by the crown. As was mentioned earlier, much of the territory had been given by Charles II to William Penn in 1681, in payment for an old royal debt to Penn's father, and all land that had not in the meantime been sold now belonged to his heirs, Thomas and Richard Penn. They were men of a very different stamp from William. William Penn had been a staunch Quaker, willing to suffer for his convictions. He was expelled from Oxford for his religious nonconformity and imprisoned for some time in the Tower of London for writing a tract against the Trinity. When awarded the territory of Pennsylvania, he tried to govern it in accordance with his beliefs, making it a colony where religious and political freedom would flourish. Not so Thomas and Richard. They were self-seeking absentee landlords, living in London, Quakerism disavowed, with no sense of responsibility toward the Pennsylvanians who bore their family name. Nothing in their view of the world could conceivably prepare them for the idea that they might be subject to taxation by their remote tenants and neighbors. The latter, of course, were unwilling to bear the local military tax burden alone.

In 1757 the Pennsylvania assembly voted to send Ben Franklin to London as its agent, to petition the crown against the recalcitrant Penns. The petition was viewed as reasonable, in accordance with basic principles of English law, and favorable action was anticipated once the authorities in London were made aware of the problem. There was never any hint or thought of rebellion—on the contrary, Franklin often spoke of the "impossibility" of such an event. Franklin was accompanied by his son William (age twenty-six), who earned a law degree in London, and by two servants. They moved into lodgings with a widow, Margaret Stevenson, at 7 Craven Street, close to the Strand, and hired their own coach for twelve guineas a month. The Craven Street house remained Franklin's residence in London through both his missions, and is still there today, essentially unchanged (Fig. 3). There is an attractive tavern of the same period, the *Ship and Shovel*, around the corner. Whether the Franklins frequented it is not known.

Franklin's mission can be described as a partial success for the colonists. The principle that the proprietors were taxable was established, though a compromise had to be made that limited the money to be paid on this occasion to a very small amount. For Franklin personally the mission must have been disappointing. The impropriety of direct dealings between Philadelphia and London (instead of through the crown's official Governor) was the chief concern of official London. Franklin was snubbed, treated as a lowly messenger. When the Privy council was ready to respond to the petition (in November 1758, sixteen months after Franklin's arrival in London), they sent their reply directly to the Assembly through the Governor.

But Franklin had a resilient spirit. There is no evidence to suggest that indifference or hostility from official London had any effect on his self-confidence or cheerfulness. He was known and respected in intellectual circles as a scientist and philosopher. That was more important than what they thought of him in government offices.

Franklin and his son traveled a great deal and were welcomed hospitably wherever they went. They were at Cambridge University in the middle of 1758, and from there visited Northamptonshire, the original home of the Franklin family. They went to Scotland later in the year and stayed for six weeks. They met Adam Smith there, the famous economist, and David Hume, the philosopher, and Dr. John Pringle, who subsequently became one of Franklin's closest friends. Franklin received an honorary doctor's degree from St. Andrews University, and from that time on was always called "Dr. Franklin," at least in Europe.

Franklin's mission in London was extended by the Assembly (he was asked to act for them on some financial matters), and he and William visited Belgium and Holland in the summer of 1761, including a trip to Leiden, where they met Pieter van

Figure 3. Craven Street in London today (left). Benjamin Franklin lived in the house just beyond the Waldor Hotel. A plaque (above) commemorates the fact. The house was owned by a widow, Margaret Stevenson, and Franklin had four rooms on an upper floor. These photographs were taken by the author in 1985.

Musschenbroek, the inventor of the "wonderful bottle" so essential to Franklin's electrical experiments.

There is no doubt that these were happy years for Franklin and that he especially enjoyed Scotland. "The time we spent there was six weeks of the densest happiness I have met with," he wrote to Lord and Lady Kames after his return from Scotland. Were it not for other obligations, "Scotland would be the country I should choose to spend the remainder of my days."

One of the attractions of all of Western Europe was a much more widespread interest in scientific matters than existed in America. Franklin had almost no time for science in those last few years in Pennsylvania that preceded his London mission. He would have had little encouragement had time been available. In England there were both time and encouragement.

Franklin's avid scientific curiosity, dormant when he was fully occupied by affairs of state, revived the very first day he was in England. His ship landed in Falmouth, in the southwest corner of England, and he immediately set out for London overland by stage, stopping on the way "to view Stonehenge on Salisbury Plain," then as now an enigmatic scientific puzzle. A year later he was in Cambridge, doing a joint experiment with Professor John Hadley, on cooling by evaporation. The wave-stilling experiment done during Franklin's second stay in London was not an isolated incident, but typical of his normal pattern of enlightened life in England.

Franklin returned to Philadelphia in November 1762, but did not remain long. Establishment of the principle of the Proprietors' tax liability had done little to assuage discontent. Very little money could actually be collected, and the Proprietors' arrogance and insensitivity to the needs of their territory's actual inhabitants continued. The Assembly therefore took a more serious step, a petition to the crown to remove the Proprietors from power altogether and to change the status of Pennsylvania to that of a colony under direct royal government. This petition did not have undivided support. The city of Philadelphia in particular had many voters with vested economic or social interest in continuing the status quo, and Franklin and some of his friends were defeated for election to the Assembly in 1764. Nevertheless, the "Old Party" won overall and at once sent Franklin on his second mission to London, to present and defend the controversial petition, which, as it turned out, was actually never discussed or acted upon.

4. Ten Tempestuous Years

It is difficult to imagine that a highly respected patriot, in his sixties, could overnight become an object of vicious calumny and public disgrace; labeled Judas, traitor, and thief; an angry mob ready to burn down his home. It is even more difficult to imagine that a man so savagely abused should be able to retain his equanimity and continue his political life as constructively as heretofore, with wisdom and wit and determination, and (at least publicly) without rancor. Yet Ben Franklin did just that. He was twice in disgrace, first with his fellow citizens in Philadelphia in 1765, when he appeared to have sided with Britain in its imposition of the Stamp Act on the American Colonies. "A firm loyalty to the Crown . . . is the wisest course," he had said, after fruitless attempts to prevent enactment, and he tried to moderate the resentment against the new taxation by persuading Britain to appoint local Americans as administrators. At home they thought he had been bought, perhaps by royal favors.

The second occasion was in 1774 in London, when he was insulted and berated at a Privy Council hearing that was ostensibly called to consider a Massachusetts petition.

It was in the quiet period between these two storms that his experiments with oil spread on water were carried out. Perhaps they helped to calm the man as well as the waves.

Franklin's first fall from public favor was dramatically reversed in 1766, when the Stamp Act was repealed. Franklin was given credit (probably more than he actually merited) and transformed from goat to hero. Appointments as agent for Massachusetts, Georgia, and New Jersey were added to his similar position for Pennsylvania. He became de facto spokesman for all of America, an ambassador before America had the legal right to send one. He could no longer be snubbed in London and began to have access to the Prime Minister and to Parliament.

This period was again a happy time for Franklin, perhaps as much as the years of his first London mission. He occupied the same comfortable lodgings in Craven Street as before. At least initially his faith in a harmonious solution to British-American disputes persisted. For a long time he continued to believe that George III was an upright and virtuous king (unlike George I and II he could speak English!), instead of the villain of the piece, which he ultimately turned out to be. In the summer of 1766 Franklin toured Holland and Germany with his friend Sir John Pringle, and in August the two of them went to Paris together. There were frequent visits to Jonathan Shipley, Bishop of St. Asaph, in Hampshire, near Winchester, and Franklin began his unfinished autobiography there in 1771. In the same year there was a long journey to Scotland and Ireland, including one month in David Hume's home in Edinburgh. In 1772 came the trip to the Lake District and other northern parts of England, during which the effect of oil on water was demonstrated on at least two separate occasions. Travel in those days was far from luxurious (see Boswell's journal for his 1773 trip with Dr. Johnson to the Hebrides), but Franklin's constitution was strong:

> In Cumberland I ascended a very high mountain where I had a prospect of a most beautiful country, of hills, fields, lakes, villas, etc., and at Whitehaven went down to the coal mines till they told me I was eighty fathoms under the surface of the sea which rolled over our heads; so that I have been nearer both the upper and lower regions than ever in my life before.

Franklin's subsequent disastrous fall from public favor in London had a complex origin, not directly related to his mission. In 1768 and 1769 Thomas Hutchinson, native of Boston, shortly to be appointed the crown's governor of Massachusetts, and his second in command, Andrew Oliver, had written letters to Thomas Whately, private secretary to Prime Minister Grenville in London, urging drastic action to cope with the spirit of rebellion in the colony. Armed forces were recommended. "There must be an abridgment of what are called English liberties," Hutchinson said in one of his letters. In 1772 (both Grenville and Whately had died in the interim) the letters somehow fell into Franklin's hands. He deemed it his duty to make their contents known to the people of Massachusetts, so that they should understand that hateful taxes and excesses by

British soldiers originated with one of their own native sons. Don't blame the English for everything, was what he wanted to communicate.

The Massachusetts Assembly then humbly petitioned the King, through their agent Franklin, to remove Hutchinson and Oliver from office. They were the ones responsible for disturbing the harmony between London and the colonies, they wrote, the cause of all the "misery and bloodshed." The letters were cited as evidence, and this caused the same kind of commotion that might occur today when official security is breached. Where was the leak? Who was the thief? Was it Whately before he died? A witch-hunt was in progress, a man called John Temple was wrongly accused, and a duel took place between Temple and Whately's brother William. Franklin was alarmed and wrote an open letter to the newspaper *Public Advertiser*, declaring that he alone was responsible for sending the letters, giving his logical reasons for doing so, but not revealing who gave the letters to him.

The Massachusetts petition for removal of Hutchinson and Oliver was now brought before the Privy Council with dispatch, in January 1774. The hearing took place in the "Cockpit," so named because the council chamber was on the site in Whitehall where Henry VIII had built a cockpit (i.e., a place for cockfighting) and a tennis court next door. When a new building was erected on the site the old name was preserved: until the nineteenth century letters from the council chamber were formally headed "The Cockpit." Franklin, as Massachusetts agent, was required to be present at the hearing, but with no premonition of what was to come; that it would not be a discussion of the political issues at all, but an attack on Franklin's character, integrity, and motives.

England's solicitor general (Alexander Wedderburn) was retained as Hutchinson's counsel, and he directed the abusive tirade against Franklin. The Privy Council was present in full force, including the Archbishop of Canterbury, the Duke of Queensberry, numerous peers of the realm. They listened gleefully to Wedderburn's invective. The spectators laughed and applauded as the old gentleman before them was branded a common thief, accused of stealing the letters because he himself wished to become Governor of Massachusetts.

> I hope, my Lords, you will mark and brand the man for the honour of this country, of Europe, and of mankind. . . . He has forfeited all the respect of societies and of men.

Wedderburn's harangue lasted an hour. Joseph Priestley was among the spectators, probably the only true friend of Franklin's in the room. Franklin himself stood in silence throughout the attack (only the Privy Councillors had seats), and declined to make a rebuttal statement.

The Massachusetts petition was of course dismissed as groundless. The following day Franklin was removed from the position of deputy postmaster-general, which he had held for twenty years, and this action perhaps hurt him more than Wedderburn's tirade because he had made the post office a profitable operation for the crown, and was proud of it. Franklin was disgraced, his usefulness as agent for American interests ended. It is interesting to note, however, that disgrace did not stretch to scientific circles. Franklin's detailed account of his oil-spreading experiments was read before the Royal Society in June 1774 while the Privy Council meeting must still have been fresh in everyone's mind, and it was published in the *Philosophical Transactions* shortly thereafter.

Franklin returned home in 1775 on the Pennsylvania packet. As usual, his inquisitive mind found useful occupation—charting the edge of the Gulf Stream by lowering a thermometer into the water several times a day. He noted that ship captains could profitably employ a thermometer as a tool to speed up their voyage, staying out of the Gulf Stream on the way west and keeping within it on the return.

He was elected to the Continental Congress the very next day after his arrival in Philadelphia, becoming its oldest member. A year later he was the oldest among those who proudly and courageously signed the Declaration of Independence. It is usually the young who are in the forefront of revolution, and Franklin was young in spirit if not in years. Sadly for Franklin, his own son William, no longer young at forty-five, remained a loyalist. Ben Franklin prized family harmony, and his grandson (illegitimate child of William) became very close to him at this time.

Franklin, now seventy, was soon on his way again. In October 1776, accompanied by his grandson, he went to France for another nine years of expatriate living, this time as an official representative of "The United States of America." The mission was to obtain French financial support for the now inevitable war between America and England. The journey, on the warship *Reprisal*, was perilous, for Franklin would surely have been hanged as a traitor if caught by the English. There is no record of scientific research on this journey, but there was minor military action: the *Reprisal* captured two English brigantines en route.

Ben Franklin remained in France until he was succeeded by Thomas Jefferson in 1785. He was urged by his friends to use the leisure afforded by the return journey to write extensive memoirs of his years in France, for undoubtedly no foreigner knew France as well as he did. He did not do that, but used the time instead for science, writing a pamphlet on "Cause and Cure of Smoky Chimneys."

Ben Franklin died in Philadelphia on April 17, 1790.

CHAPTER THREE

Friends and Influences

1. The London Scene

JAMES Boswell is the great chronicler of London life. His journals are vivid in content, sparkling in language. They record every event, every conversation, every promenade along the Strand, and every one of his numerous (often disastrous) sexual escapades. Shamefully embarrassing to his strict Presbyterian Scottish family, the journals were suppressed and hidden away in a castle in Ireland. They were not discovered and revealed to the world until the 1920s.

Boswell was history's most literary hedonist, Samuel Johnson's biographer and traveling companion, and, above all, a passionate cultivator of every celebrity of the day, a source of accurate descriptions and biting comments about dress and speech and attitudes. He mentions "Dr. Franklin"—always thus, without first name—several times. He had Franklin to dinner with Sir John Pringle in May 1768. He found Franklin playing chess with Pringle when he visited the latter in September 1769. "Sir John, though a most worthy man," he wrote, "has a peculiar, sour manner. Franklin again is all jollity and pleasantry."

But Franklin doesn't reciprocate, doesn't mention James Boswell in any of his letters. Boswell may have been too vainglorious for Franklin's tastes, and too indiscriminate in his choice of friends. Besides, Boswell had little sympathy for the American cause, which was about practical issues of trade and taxes, much less romantic than the cause of Corsica's struggles for freedom from Genoese domination, about which Boswell wrote two books. There is no reason why Franklin should have found his meetings with Boswell particularly memorable.

There was another glossy segment of London society in which Franklin would have had greater interest, because it would have furthered his mission to do so. This was the fashionable world that revolved around the king's court and political or ecclesiastical circles, but it was a world to which Franklin could have no access. He was not an "ambassador," was never presented to King George, did not associate with other ambassadors. He was not entertained at the houses of cabinet ministers or other influential members of society. America was not a country; it was a colony, a remote outpost of Britain, as remote as the distant Scottish isles, where barbarian clan chiefs still ruled their primitive crofters. No one would have denied that there were men of stature in America, brave men who cleared the land and repelled the onslaughts of Indian savages for the greater glory of king and country. But England had governors in America, official representatives of the King, and any legitimate American visitor would have had at least a governor's introduction to the court or, even better, support for the cause he was pleading. Ben Franklin, this plainly dressed petitioner from the "Assembly" of Pennsylvania (whoever they might be), had no official status.

In other times, as scientist and philosopher, he might have found a congenial fellowship at the universities in Oxford or Cambridge, though he lacked the formal credentials that would have made him automatically acceptable. Franklin visited Cambridge twice in 1758, with great enjoyment, but the middle of the eighteenth century was on the whole a period of stagnation for the English universities. The Dissenting Academies and the Scottish universities were doing much more to nourish culture than Oxford and Cambridge. G. M. Trevelyan, writing about Trinity College, Cambridge, calls it "unworthy of its past and its future" during this period. (See Fig. 4.)

Who then were Franklin's friends during sixteen years in London, without a wife, probably without even a mistress to divert

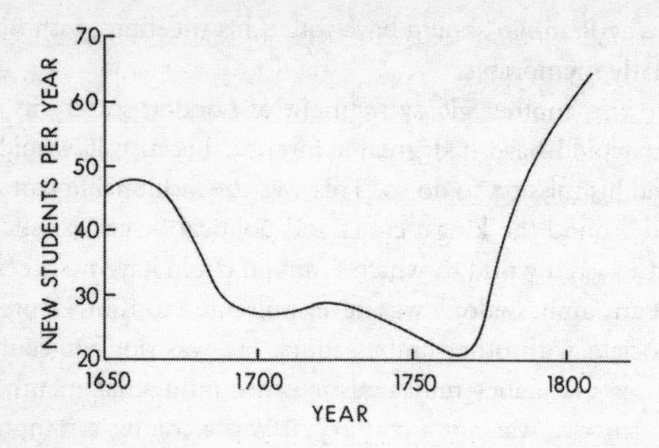

Figure 4. Annual admission of new students to Trinity College, Cambridge University, based on data from G. M. Trevelyan's historical sketch of Trinity College. After 1900 admissions rose to more than two hundred a year. Throughout the period covered by the graph, Trinity admissions represented about one-sixth of those for all of Cambridge. In the words of Trevelyan: "The decadence in the eighteenth century . . . is equally marked in Trinity and in the aggregate of the other Colleges."

him from serious pursuits? One of them was William Strahan, master printer, who had printed Samuel Johnson's dictionary, and was later to publish the works of Gibbon, Adam Smith, and others. He and Franklin had exchanged correspondence for fourteen years before Franklin's first London mission. Another was Richard Jackson, English lawyer and subsequently Member of Parliament, who was a secret supporter of the American cause. Bishop Shipley and his family in Hampshire were friends in the country, far from London. Lord Henry Kames was a friend in Scotland—a judge and philosopher who wrote books on criticism and "the Art of Thinking."

Most of Franklin's friends, however, were scientists or philosophers, religious dissidents, intellectuals, members of discussion groups like his own Library Company in Philadelphia. It is important to give some account of them here, of the way they lived and conducted their scientific work. They were the people with whom Franklin discussed his ideas, by letter before he lived in London, and in person afterward. They were the people who influenced Franklin, if anyone did. Most of his scientific friends, in fact most of the scientists of the period whom we remember today, were amateurs, like Franklin himself. They lacked university degrees or other formal credentials. Many of them had received no formal scientific training of any kind. Even at the Royal Society, the official temple of scientific wisdom, few of the Fellows had university affiliations or even degrees. Apart from being a "colonial," Franklin was not greatly different from the rest of the Royal Society membership.

With respect to intellectual life, there was a considerable difference between England and the rest of Europe. In particular, the scholarly professions tended to be more structured on the European Continent, more like they are today. Johann Carl Wilcke (1732–1796), for example, was required to make a formal choice between following his father into the ministry or embarking on a scientific career—he could not, as he would have done in England, do both. He chose science, which required an arduous course of study at the University of Rostock and a proper doctoral dissertation, which had to be defended before a committee, much as we do nowadays. Wilcke passed the requirements with honor and some fame in 1757 and subsequently became a professor in Sweden.

There is some irony in this, because Wilcke's thesis was on the subject of Ben Franklin's electrical experiments of a few years earlier. Its most important aspect was a critical examination of the objections to Franklin's work that had been raised by the French electrician, the Abbé Nollet—which Wilcke was by and large able to discount. It was a good dissertation. The structured society was clearly able to produce elegant scholarship. But the big scientific advance, which provided the motivation for the scholarship, had come from a self-taught man in the unstructured (virtually nonexistent) scientific environment of Philadelphia.

2. The Club of Honest Whigs

The Royal Society Club (founded in 1743) was the largest and most prestigious of the eighteenth-century discussion and dining clubs that met in taverns or coffee houses, and it was also the most formal. It met at the famous Mitre Tavern on Fleet Street, where Samuel Johnson regularly held court. Only Fellows of the Royal Society were admitted to membership in the club, and politics and religion were strictly forbidden subjects for discussion. Careful records of attendance were kept and are preserved, and from them we learn that Franklin was frequently invited—as well as the revealing fact that his invitations became more numerous after the Cockpit hearing, where official England had declared him unfit for association with decent men. Franklin, as a Royal Society Fellow, could of course have become a member of the club in his own right, but did not join, being always convinced that he would return home very soon. When present, he was always the guest of another member.

The Club of Honest Whigs was less formal. As the name "Whig" implies, this club encouraged discussion of politics and religion as well as scientific matters. Nonconformist clergymen were among the most regular members. Joseph Priestley, Sir John Pringle, Peter Collinson, Richard Price, John Fothergill—Franklin's most cherished London acquaintances—were all members.

James Boswell was apparently a member in 1769. In his journal for September 21 he wrote:

> I went to a club to which I belong. It meets every other Thursday at St. Paul's coffee house. It consists of clergymen, physicians, and several other professions. There are of it: Dr. Franklin, Rose of Chiswick, Mr. Price, who writes on morals. . . . We have wine and punch upon the table. Some of us smoke a pipe, conversation goes on pretty formally, sometimes sensibly, and sometimes furiously. At nine there is a sideboard with Welsh rabbits, apple puffs, porter and beer.

> Our reckoning is about 18 pence a head. Much was said this night against Parliament.

Since Boswell was a staunch Tory and is not noted for original contributions to philosophical thought, he would appear to be out of his element at the Club of Honest Whigs, the conversation usually well over his head, unless they were polite when he was there and talked only about Corsica, which is unlikely. As a matter of fact, there is every evidence that he was a rare attender—perhaps years passed between visits. In 1772, when the coffee-house had moved from St. Paul's Churchyard to nearby Ludgate Hill, becoming the London Coffee House, and the club had moved with it, Boswell went there to meet Franklin. "When I got thither," he writes, "neither he nor any one else whom I knew was there. One of the men here came out, but could recollect nothing of me. This was very awkward." Eventually someone recognized him, he was admitted, and he found Priestley there, whom he knew.

For Franklin the Club of Honest Whigs was a necessity, an oasis of friendship in an often hostile city. Years later, after the American Revolution, he wrote from France to Richard Price that his best memories were of the London club: "I often think of the agreeable evenings I used to pass with that excellent collection of good men."

3. Biographical Sketches

Today science is a career, with good salary and fringe benefits. Intellectually stimulating? Perhaps for a few, but for most it's just a way of earning a secure living, five days a week, with the weekend reserved for more exciting activities. It was different in the eighteenth century, which I try to illustrate here with some brief biographies. We see men eager to learn, to use their imaginations, unafraid to challenge established dogma—men who worked at science because they wanted to, because they found it exciting.

We should also allow our minds to wander a little as we read the biographies and to note some peripherals. One is that none of our cast of characters was a mathematician, which is a circumstance we shall consider again in the following chapter. A second peripheral is evidence for a remarkable spirit of tolerance in eighteenth-century England—Sir John Pringle's continued employment as royal physician in spite of his friendship with the traitorous Dr. Franklin; the flourishing of "dissenting academies," co-existing with the universities at Oxford and Cambridge where religious conformity was rigidly enforced; the country's permissiveness toward Priestley's outrageously provocative religious views. It was not till after the French Revolution that fear began to create some repression of independent thought or writing.*

Peter Collinson (1694–1768). Peter Collinson was older than Franklin, very active in the English Quaker movement, but perhaps no longer active as a scientist even during Franklin's first London mission. He had, however, been Franklin's patron for twenty years, and had sent to Philadelphia the German glass tube (with instructions) that started Franklin on his electrical experiments in 1745. Collinson was responsible for publication of Franklin's electrical studies in London, at first letter by letter, as they arrived, and then in 1751 in book form. He sponsored Franklin for membership in the Royal Society, and for the Society's Copley Medal. Franklin and his son were guests in Collinson's house in Mill Hill when they first arrived in London on July 26, 1757.

Collinson earned his living as a businessman, a shopkeeper, and (fortunately for Franklin) specialized in trade with the American Colonies. He was a collector of curiosities and especially interested in gardens and plants. He prevailed upon his American merchant

*Americans may find this tolerance at home inconsistent with England's attitude toward the Colonies, but I don't share that view. Even in America you were free to *say* what you wanted, provided that you paid your bills and taxes. Remember that England's penalty for not paying bills was confinement to debtor's prison, and that the penalty for theft was hanging!

friends to send him seeds as well as orders for goods, and he became famous for his ability to cultivate American plants in the English climate, especially at his house on Ridge Road in Mill Hill, now part of Mill Hill School, where he had an extensive garden, containing what to English eyes was an exotic assortment of flowers. The most famous of Collinson's business associates was John Bartram, whom George III subsequently appointed as American botanist to his court. Part of the original Bartram farm still exists as part of Philadelphia's Fairmount Park system. A sign proclaims it to be "America's oldest surviving botanic garden and first commercial nursery."

Collinson was an active member of the Royal Society, but produced little original scientific work. His best-remembered paper (published in 1760) concerned the migration of swallows. It consisted of a well-reasoned argument, based on correspondence and travelers' accounts, that swallows migrate between summer and winter ranges, and that they do not, as was then commonly believed, hibernate in their summer quarters *under water*. It seems incredible that such a bizarre idea should have remained in vogue as late as 1760, but in fact Collinson had difficulty in persuading even the great Swedish taxonomist Linnaeus that the belief was invalid.

Collinson's fame and acceptance by the Royal Society rested not on this or other scientific studies, but on his self-created role as a publicizer and lubricator of the wheels of communication. For over thirty years Collinson was the center of a network of scientific correspondence that reached to Peking in the east and (of course) to Philadelphia in the west. He informed his correspondents abroad of scientific developments at home and in turn publicized their work in London, and, when sufficiently impressed, nominated them to membership in the Royal Society. He had continuous contact with the Library Company of Philadelphia and chose books and instruments to send to them, but he did the same for other correspondents in his worldwide network.

William Watson (1715–1787). William Watson was another selfmade man, but more motivated by personal ambition than

Collinson. He was a charter member of the Royal Society Club and also active in the Club of Honest Whigs. It was Watson, not Peter Collinson, who read Ben Franklin's *first* electrical letter to the Royal Society in 1748 and who sponsored its publication in the *Philosophical Transactions*. Since Collinson was not himself an electrician, it was the logical thing for him to do, to have it reviewed by the Royal Society member with greatest expertise in the field, which Watson in that particular year could claim to be, though his ultimate eminence came as a physician and not as a scientist.

Watson's somewhat colorful career began with his apprenticeship to an apothecary at age sixteen. Apothecaries at that time had the right to prescribe as well as to compound medicines, and an apprentice in the trade was exposed to more hard science than one might think. Watson became particularly interested in botany and, though he set up his own apothecary shop and earned his livelihood that way, made original contributions to botanical knowledge and became an active member of the Royal Society. He was soon diverted from botany to the more popular field of electricity where he did some innovative work that earned him the Copley Medal of the Royal Society in 1745 and led to his involvement with Franklin.

As it happened, Watson's theoretical ideas about electricity (supporting the now-forgotten French two-fluid theory) became unattractive as a result of Franklin's studies, and Watson lost interest in the field. At the same time his reputation as a "medical man" increased, his pharmacy prospered, and he sought to make medicine his formally accepted profession. To do this he first had to dissociate himself from the Society of Apothecaries, because physicians resented the apothecaries' right to prescribe medication and would under no circumstances admit an apothecary into their ranks. (It appears that a "resignation fee" was charged, which in Watson's case amounted to the huge sum of 50 pounds!) After that, Watson had to obtain a medical degree, which, according to some authorities, he "purchased" from the University of Halle in 1757. He received his medical license in 1759, became principal physician at the Foundling

Hospital in London, a position he filled with dedication and generosity. He was knighted in 1786, shortly before his death.

John Canton (1718–1772). John Canton, one of the regulars at the Club of Honest Whigs, has been described as "one of the most distinguished of the group of self-made, self-educated men who were the best representatives of English physics in the mid-eighteenth century." He was the first man in England to confirm Franklin's postulated identity of lightning and electricity, and his paper on the subject forms part of the book version of Franklin's experiments and observations that was put together by Peter Collinson.

Like Ben Franklin, John Canton received little formal education. His father removed him from school just as he was learning the rudiments of algebra, in order to get him into the family business of broadcloth weaving, and even refused to allow him a candle at night to pursue his studies on his own time. With the help of Henry Miles, a "dissenting" minister (like Priestley), he was able to escape from his father's influence and from 1738 to his death in 1772 earned his living as a schoolteacher in a private school in London. He did much research of lasting importance in electricity—apart from his work supporting Franklin's theory—and won the Royal Society Copley Medal in 1751 for his invention of a new method for making artificial magnets.

In a world that still thought "air" to be the solitary gaseous substance, unique in its ability to expand to fill all space, Canton's most important achievement was probably the discovery that water is compressible. This discovery was one of the first scientific advances that depended on the very careful determination of minute changes in some measurable quantity (in this case volume as a function of pressure), and the conclusion was so revolutionary that it was challenged within the Royal Society. A committee was set up to examine Canton's claim, and it included Ben Franklin, Charles Cavendish, and William Watson as members. The committee verified Canton's experiments and endorsed his conclusions. Canton received a second Copley Medal for this work in 1765.

Sir John Pringle (1707–1782). Sir John Pringle is an anomaly. Unlike Franklin himself and unlike Collinson, Watson, Priestley, etc., he was a member of what we would today call the upper class, the son of gentry, formally educated at the best institutions (St. Andrews and Edinburgh, with an M.D. from Leiden), and he was throughout his life a conservative member of established society. He was appointed physician to Queen Charlotte in 1761, created a baronet in 1766, and appointed physician to George III in 1774. He was President of the Royal Society from 1772 to 1778, which period includes the year of publication of Franklin's letter about oil on water.

In spite of the difference in social background, Pringle was clearly one of Franklin's closest friends. They traveled together to the Continent in 1766. Pringle was with Franklin at Derwent Water when the spreading experiment was repeated there (the occasion that the Reverend Mr. Farish heard about) in 1772. Pringle and Franklin were together in London in their encounters with Boswell in 1768 and 1769, as mentioned earlier. Franklin was a guest in Pringle's house during his last week in England in 1775.

Pringle was Scottish, which may account for his own lack of class consciousness. His wife died after only a few years of marriage and he never remarried, which may have created a natural personal affinity between him and Franklin. But how could he continue to receive royal patronage in England in spite of his close association with a man who had been branded a traitor by the king's own Privy Council?*

*Pringle eventually did fall out of favor with the King in 1777. It was over the seemingly trivial question of whether lightning rods should have pointed or rounded ends. Franklin had been an advocate of pointed ends, but blunt-enders believed that pointed ends might attract lightning that would never by itself have struck at all. George III personally decided in favor of rounded ends, but Pringle (President of the Royal Society at the time) would not support the decision. It was believed that his contrary opinion was influenced by Franklin, and it became a serious political matter because one of the uses of lightning rods was to protect stores of gunpowder from unpremeditated explosion during storms.

Professionally, Pringle's great achievements were in the field of military medicine. He was chief physician for the British army in Flanders in the 1740s. He originated the terms "septic" and "antiseptic" and was the first to lay down practical rules for military hygiene, receiving the Copley Medal of the Royal Society for this work in 1750. Pringle was also the first person to propose the idea that military hospitals should be sanctuaries from enemy action, an idea that was immediately adopted in Europe. In 1752 he published his famous book *Observations on Diseases of the Army*, which went through numerous editions (including a posthumous American one) and was translated into several foreign languages. He is generally regarded today as the founder of modern military medicine.

Josiah Wedgwood (1730–1795). Josiah Wedgwood is one of the most celebrated pottery designers and manufacturers of all time, and his descendants today still run the pottery factories in Staffordshire, not far from where Josiah was born. Josiah Wedgwood was an active member of the Lunar Society, a Birmingham analogue of the Club of Honest Whigs, founded by Erasmus Darwin (Charles Darwin's grandfather) and others in 1766. James Watt, inventor of the steam engine, was a Lunar Society member, and so was Joseph Priestly until he moved to the south of England in 1773. Wedgwood supplied Priestley with crucibles and other ceramic ware that he needed for the experimental work he did later while in Lord Shelburne's employ. Wedgwood was a member of the Royal Society because he took a scientific approach to pottery-making, and in the course of this invented a pyrometer for measuring the high temperatures in firing ovens, which was enough to earn him a Royal Society Fellowship.

Wedgwood illustrates the close ties among the small number of actual scientists in England at the time. Applied science was as respected as pure science, north as good as south. The small scientific community was one big family, and in Wedgwood's case this was true in more senses than one. Wedgwood's daughter Suzannah married Erasmus Darwin's son and eventually became the mother of Charles Darwin. Charles Darwin's subsequent wife, Emma, was his first cousin, also a granddaughter of Josiah Wedgwood.

Joseph Priestley (1733–1804). Joseph Priestley, the discoverer of oxygen, was a dominant scientific personality in his time and remains today a key figure in science history. He was probably Ben Franklin's closest associate at the end of his stay in London, when negotiations with the government had broken down. Priestley stood at Franklin's side during the infamous "Cockpit" proceedings of the Privy Council and breakfasted with Franklin at Craven Street the following morning. A year later, Franklin spent his last day in London alone with Priestley before embarking for Philadelphia, and even shed tears in his presence for the era that had ended.

The reason for their friendship is not entirely clear. Priestley was twenty-seven years younger than Franklin. He was a provocative non-conformist in religion (one of the founders of Unitarianism) and as heretical as Franklin in his politics. But he may have been partially motivated by some stubborn inner need to take unpopular positions for their own sake and not so much by common sense as Franklin was. His scientific opinions, even in the interpretation of his own experiments, seem today to have been deliberately controversial, sometimes (as in his continued belief in the phlogiston theory) against all common sense and reason. He must have been a difficult man, never "all jollity and pleasantry," as Boswell described Ben Franklin.

Priestley was the son of a cloth dresser, born in a small town near Leeds in 1733. He was educated at a "dissenting academy" at Daventry, where he was formally instructed in a then famous (or notorious) method of free inquiry into "controversial subjects" devised by a man named Philip Doddridge. Priestley held two positions as preacher in small churches, but was unsuccessful in the pulpit, partly on account of an inherited speech impediment and partly because of his heterodox views on the Trinity. He thereupon became a teacher, at another "dissenting academy," and there began his lifelong prolific productivity as a writer of books, pamphlets, and other publications. One of the books, still often cited by modern scholars, is entitled *A Course of Lectures on the Theory of Language*. In 1764 he received an LL.D. from the University of Edinburgh in recognition of his work in education, and at about the same time he began his scientific career with what proved to be an influential critical work, *The History and Present State of Electricity*. In the course of writing this book he consulted authorities on the subject, including John Canton and Benjamin Franklin, whom he first met in London in 1765. They were sufficiently impressed with Priestley to secure his election to the Royal Society in 1766.

Priestley was a poor man, with no skilled trade or business, and in 1773 entered the service of William Petty, Earl of Shelburne (later Marquis of Lansdowne), formally as librarian and "resident intellectual," but perhaps also as political aide. Shelburne provided a laboratory for Priestley, and most of Priestley's best experimental work, summarized in the famous book *Experiments and Observations on Different Kinds of Air*, was done during this period. The work included the discovery in 1774 of a new "air" (today we would say "gas") with unexpected properties, obtained by heating mercuric oxide, and this new substance of course subsequently proved to be oxygen and to herald the beginning of a revolution in chemistry as profound as the revolution in physics that Newton had initiated a century before. Priestley's single visit to the European Continent also took place while he was in Shelburne's service. He met his subsequent rival, Antoine Lavoisier, during this trip and demonstrated to him the release of the new gas from mercuric oxide to him. (There is a parallel here with Michael Faraday, about fifty years later, who also was able to afford a visit to Europe only in a "household servant" capacity with Sir Humphry Davy and his wife.)

Priestley left Shelburne's service in 1780, returning to the ministry in Birmingham, where he began to lose friends and influence among scientists because of his stubbornness and the sometimes exasperating lack of flexibility in his thinking. In spite of the evidence from his own experiments, virtually definitive in their demonstration of *chemical* differences among various gases, Priestley continued to adhere to the idea that all gases are in fact made of the same intrinsic substance, and that the differences among them have a *physical* basis. Thus Priestley continued to designate the gas released by mercuric oxide as "dephlogistinated air," and it was Lavoisier who named it "oxygen" and thereby established the true importance of Priestley's discovery.*

At the same time Priestley's theological publications were becoming more radical—for example, his *History of Early Opinions Concerning Jesus Christ*, published in 1786. And in 1791, when he publicly supported the French Revolution (implicitly favoring an end to monarchy in Britain, too?), he went too far and provoked mob violence. The New Meeting House, where Priestley was minister, and his home and laboratory were destroyed, and his personal safety was threatened. Even the Fellows of the Royal Society shunned him after that, and Priestley was persuaded that he should move to the New World, to the United States of America. He did so in 1794, settling at an enclave of other British "refugees" at Northumberland, Pennsylvania.

Some lifelong rebels mellow in old age, but not Joseph Priestley. He hardened his defiance against the new chemistry that flowered from Lavoisier's seeds, firing off letters from Northumberland to one and all, to defend the phlogiston theory, and to deny that water could be a compound, dissociable to hydrogen and oxygen. He

*This is really an oversimplification because Lavoisier was also subject to preconceived notions. The word "oxygen" means "acid-forming" and is therefore actually a misnomer. Lavoisier gave the name to the respirable component of air because he believed that oxygen was an *essential* component of all acids. It was Priestley who proved him wrong by showing that hydrochloric acid contains no oxygen.

continued to engage in theological and political controversy. "This eldest Son of Disorder will never obtain his sought-for 'Repose' on this side of the Grave," is what one of his opponents wrote, "and the Government of Heaven itself, should he ever get there, will, in his opinion, want reformation."

In his final years Priestley was befriended by United States President Thomas Jefferson and received some solace from being for a change "in favor" in high places. Priestley died in Northumberland in 1804 at age seventy-one.

CHAPTER FOUR

The French Connection

France is different today, was different then. Its special charms have somehow survived unchanged through the reigns of profligate kings, through violent revolution and devastating wars. Franklin, on his first visit in 1767, noted the high fashion of Parisian women, the rows of trees lining country roads, the high quality of the wine, the same as a tourist today might. In 1779, when he lived as America's plenipotentiary on the grounds of a great estate in Passy, on the western side of Paris, he had 1040 bottles of wine in his cellar. In September 1782 the inventory was 1203 bottles.

Above all, he loved the French ladies, and, if reports of the period are to be believed, they loved him too, despite his age and homely dress and behavior. "The French ladies," he wrote to his stepniece in Boston, "have . . . a thousand . . . ways of rendering themselves agreeable." He never wrote anything like that from England! Future American President John Adams and his wife Abigail, visiting Franklin in Paris in 1784, did not approve. Their puritanical New England upbringing did not prepare them for the casual displays of affection between Madame Helvétius (who had been invited to dinner) and the venerable and venerated Dr. Franklin.

France was uniquely different from the rest of the world in its attitude toward science. In the reign of Louis XIV it had become a country of elegance and extravagance, exemplified for us today by the palace at Versailles. Louis XV, who succeeded him, was King of France throughout the period of Franklin's scientific work, and he continued the traditions of Louis XIV, with less restraint and some predisposition toward vulgarity. The pursuit of science was encouraged—the Enlightenment was not just an intellectual movement, but part of the fashionable world at court. More than that, science was able to contribute to the glory of the court, providing embellishments of a kind to which England's dull Hanoverian monarchs never dreamed of aspiring. An offshoot of botanical studies, for example, was the provision of exotic plants for the Royal Gardens, and distinguished botanists were always to be found on the royal payroll.

Electricity was a branch of science that particularly delighted the King. He wanted grand spectacles when he entertained, and he got them from his resident electrician, the Abbé Nollet, who could use electric shocks to make gendarmes and Carthusian monks alike involuntarily leap into the air in unison, to the boisterous delight of the King's guests. The Abbé Nollet (who perhaps wasn't really an "abbé") was, however, an active experimenter and natural philosopher as well as an entertainer, with his own laboratory in Versailles and with his own doctrine as to the nature of electricity. This was the doctrine of *effluence* and *affluence*, a somewhat mystical elaboration of the two-fluid theories of electricity that were popular at the time. He was also private tutor to the dauphin, and his position at court undoubtedly depended a great deal on the aura of scientific authority that he was able to promote.

Franklin's work on electricity established a more convincing single fluid concept of electricity and thereby threatened to demolish Nollet's status as a scholar. His rivals in France jumped at the opportunity this gave them to displace the Abbé from royal favor. At the urging of the Comte de Buffon (director of the Jardin du Roi), Thomas François d'Alibard produced a French translation of Franklin's book on electricity within a few months of the publication of the first English edition. D'Alibard also was the first to do the experiment that Franklin suggested in his book as a test for the "sameness of lightning with electricity," but had himself not yet carried out. D'Alibard "fetched the lightning from the sky" by means of a pointed conductor in May 1752, a month or so before Franklin himself tested the theory with his celebrated kite experiment. The Abbé Nollet flew into a rage. This was all a plot devised by his enemies. These discoveries about lightning could not conceivably have come from overseas. From what libraries, what institutions, could this sort of intellectual insight derive? The supposed translation must be pure invention, the reported experiments a fraud.

This is of course just the sort of dispute, with its threat to the privileged position of an unpopular courtier, to capture public attention. All the world (including the King) eagerly awaited the outcome of successive trials of the lightning experiment, and Franklin's fame in France quickly far surpassed his fame in England. He was warmly welcomed when he first visited France with Sir John Pringle in 1767. He was made a foreign associate of the French Royal Academy in 1773, an extraordinary honor because the total number of such associates in all fields of science and philosophy was limited to eight. (Pringle, incidentally, was also a member of this select group and Joseph Priestley became one later, in 1784.)

Late in 1776 Franklin came to Paris to make a new expatriate home for himself. This time he was the official representative of the nascent American republic, and his position was very different from what it had been when he lived in England. He was formally received at court by Louis XVI (Louis XV had died in 1774) and became a cherished guest in high society. He talked several times to Voltaire and took part in the Masonic ceremony that honored Voltaire after his death in 1778. He became a popular hero. He deliberately affected plain clothes and unfashionable spectacles because that was the image the French had of him, that was how they wanted to see him.

Electricity had by then run its course as a royal entertainment. Balloons were the fashion now, and Franklin was present at the first balloon ascent in Paris in 1783. Another fad was "animal

Figure 5. "Magnetism is exposed." This cartoon shows the mesmerists in flight. Benjamin Franklin is seen at the lower left, brandishing the report of the Royal Commission. From Robert Darnton, *Mesmerism and the End of the Enlightenment in France*, Cambridge, Mass.: Harvard Univ. Press, 1968. The original cartoon is in the collections of the Bibiliothèque Nationale in Paris.

magnetism," an occult force by means of which Friedrich Anton Mesmer claimed to be able to heal the sick. The Queen (Marie Antoinette), the Marquis de Lafayette, and other nobles were among Mesmer's ardent followers. Physicians and scientists were less impressed and persuaded the King to appoint a commission to investigate Mesmer's claims. Franklin was one of its members, and some of the tests were carried out at his house. Other members were France's most famous chemist, Antoine Lavoisier, and a mild-mannered physician by the name of Dr. Guillotine, one of whose missions in life was to make executions less painful. The commission eventually rejected Mesmer's claims as fraudulent. (See fig. 5).

Franklin returned to Philadelphia in 1785 and died in 1790. He heard about the beginnings of the French Revolution, but was spared knowledge of its later excesses. His old friend Lavoisier was tried before the Revolutionary Tribunal on the morning of May 8, 1794, and convicted and executed on the same day. Appeals for clemency because of his potential value to the Republic were denied. "Nous n'avons pas besoin des savants," they said.

Enlightenment had ended, at least in France.

CHAPTER FIVE
Pliny The Elder

> I had, when a youth, read and smiled at PLINY's account of a practice among the seamen of his time, to still the waves in a storm by pouring oil into the sea; which he mentions, as well as the use made of oil by the divers.
>
> Solitary reference to a printed work in Franklin's letter to William Brownrigg

A YOUNG couple moved into the house next door about the time I was writing this chapter, and I introduced myself to my new neighbors. "I know you," the young man said, "I've used your book." It's a nice distinction, *used* rather than *read*, and I felt flattered.

Ben Franklin had read Pliny and had smiled at what he wrote, which implies he did not use him, and modern historians of science would applaud his caution. Pliny the Elder, born at Lake Como in the reign of Tiberius, was an indefatigable reader and prolific writer, who needed little sleep, had a slave read to him even during rubdowns after his bath, and had a secretary at his side for dictation at all times. His thirty-seven-volume *Natural History* consists of an outpouring of all reports about the natural world that he could find in the works of over five hundred earlier Greek and Roman writers, with added comments from his own experience. Most of it is undoubtedly shallow, misleading, often wrong. The distinguished historian Charles Singer, normally reluctant to express subjective criticism, clearly regards Pliny with distaste and calls him "scientifically far inferior to his sources."

The historical judgment may be a little too harsh. Encyclopedias can never be more than entry ports into the world of knowledge. We hope that their contents have been filtered, doors leading into false trails closed off, but this is never entirely true, even in modern encyclopedias written by massive teams of experts, and we can hardly expect it from a work by a single individual from the first century A.D. The marvel is that such a work exists at all, identifying what there is to know. Book 2 is about the physical universe—cosmology, astronomy, meteorology, geography, geology—and includes (in chapter 106) a catalog of common knowledge about the sea and rivers and springs, where the action of oil on rough seas is mentioned:

> Again everybody is aware that all springs are colder in summer than in winter, as well as of the following miracles of nature: that bronze and lead sink when in mass form, but float when flattened out into sheets; that among objects of the same weight some float and others sink; that heavy bodies are more easily moved in water; . . . that the sea is warmer in winter and saltier in summer; that all sea water is made smooth by oil, and so divers sprinkle oil on their face because it calms the rough element and carries light down with them; that on the high sea no snow falls; that though all water travels downward, springs leap upwards, and springs rise even at the roots of Etna, which is so hot that it belches out sands in a ball of flame over a space of 50 to 100 miles at a time.

Surprisingly, when Pliny speaks of objects of the same weight that might float or sink, depending on density, no mention is made of Archimedes. Admittedly, some of the "miracles of nature" are more miraculous than others.

By Pliny's count, Book 2 contains 417 individual items of this sort. Book 6 (geography) mentions 1195 towns, 576 races, and 115 famous rivers. Book 7 is about human disease. Book 8 about land animals, Books 12 to 17 are about trees. Where else could Franklin and other youths of his time have had their imaginations fired by the immensity of what there is to know? The Enlightenment rightly taught them to be skeptical, but there must first be something to be skeptical about.

Whatever his scientific shortcomings, Pliny is a memorable figure in Western history because he is one of very few scientists who died a hero's death in quest for knowledge. Pliny was in the Bay of Naples at the time of the great eruption of Vesuvius in A.D. 79, and, instead of beating a prudent retreat, rushed to shore at Pompeii to observe the eruption more closely. The story, told by his nephew Pliny the Younger, then eighteen years old, is a literary classic:

> My uncle was in charge of the Roman fleet at Misenum. About two o'clock on August 24th my mother showed me a cloud which was very big and looked odd. My uncle had spent his day as usual. He had been sunbathing, had taken a cold bath and had his lunch. He was now studying, when he heard about the cloud. He called a slave to fetch his shoes, climbed up a hill and got the best view he could of the mystery.
>
> The cloud was getting bigger and it had a flat head, like a Mediterranean Pine tree. It was carried up into the air on a very long trunk which broke into branches. The trunk was made by the blast from the volcano pushing up hard. As it got higher into the atmosphere, the blast was less strong and the ash got heavier and began to fan out. The cloud sometimes glowed white hot, sometimes the earth and ash in it made it dirty and blotchy. For a long time we did not know which mountain it was coming from. It was only later we found it was Vesuvius.
>
> My uncle, who was a great scholar, just could not keep away. . . . He ordered all the warships to be launched and he went on one himself. . . . He sailed right into the middle of the danger.
>
> As he sailed to the town, he made careful observations. He

> took notes of everything that was going on. You can see how unafraid he was. When they got near Pompeii, the ashes were hotter and fell more thickly. They were also hit by pumice and stones. These had been burned black and broken into pieces by the fires. All of a sudden they sailed into the shallows, which were full of debris from the volcano. For a moment my uncle stopped and thought he might turn back. When the helmsman said he should, he replied, "Fortune favors the fearless. Sail round to the villa of Pomponianus!"

Once there, however, they could not escape because the wind was strong and from the wrong direction.

The younger Pliny, of course, was still at Misenum, and what follows must be strongly flavored by his imagination, but it is worth-while to quote some of it:

> In every part of the world it was day. In Pompeii it was the blackest of black nights. Those who were left in the city, only got rid of the dark by torches and all sorts of lamps.
>
> My uncle decided to go out onto the shore and see for himself if the sea would let them sail. But it was still angry and against them. His slaves put down a sheet for him to lie on and he demanded and drank one or two cups of cold water. Then the flames and the smell of sulphur, which always tells you the flames are coming, made the others run away. These flames made him wake up. He stood up, leaning on two young slaves, but he fell down straight away. I suppose the thick fumes had blocked his windpipe and closed his gullet which was always weak and giving him trouble. When they found his body in the light two days later, there was not a mark on it. . . .

Pliny the Elder was fifty-six years old when he died. As with Ben Franklin, age had evidently not tempered his enthusiasm and scientific curiosity.

CHAPTER SIX

Eighteenth-Century Science

Our century is called . . . the century of philosophy par excellence.

Jean le Rond d'Alembert, 1759

1. Apologia

THE nucleus of this book is science. It's about atoms and molecules, and how molecules can organize themselves into different structures: three-dimensional crystals, raindrops, thin films, etc. In the final chapter we shall talk about molecules organizing themselves into cells that can contain and support the elements of life.

Here, as an introduction, we must ask about science in Franklin's time. What did they know about atoms and molecules? Could or should Ben Franklin and his friends have recognized the molecular significance of the spreading of oil on water? If not, what concepts or experiments were missing? When and in what connection were the missing links established?

There is a danger in looking backward. We must guard against the impulse to make judgments based on present knowledge, on what we *now* know to be right and what we *now* know to be wrong. It is easy to make fun of old-fashioned notions, but rarely is it justified. The present generation of scientists can hardly claim to be smarter than our eighteenth-century colleagues. Would we have thought differently back then, knowing only what they knew?

On the other hand, there is equally little justification for the opposite extreme of uncritical impartiality, in which *all* present perception of missed opportunity or mental error is ascribed to hindsight. Look at the twentieth-century scientific scene around us! There are certainly fools out there and imposters and charlatans. And they can sometimes attract ardent followers for a while—true believers in health food fads, for example—because we are a tolerant society with a liberal attitude toward new ideas. We believe that false ideas will die automatically from the eventual weight of evidence against them. We tend to be suspicious of authoritarian nipping in the bud, recognizing the natural tendency of "authority" to protect vested interests.

The twentieth century has no monopoly on folly, and tolerance was surely at least as prevalent during the Englightenment as it is today. So we must imagine ourselves to have been there in person, meeting at a coffee-house or tavern with like-minded friends, to discuss Newton versus Descartes, phlogiston versus oxygen, and other philosophical controversies. We should make an effort, in what follows, to distinguish good from bad, reasonable wrong turns in science from "gobbledygook"—aware that our judgment may not always be right.

2. Newton's Legacy

Isaac Newton died in 1727, the span of his life overlapping with Ben Franklin's. Franklin had actually hoped to meet Newton when he first came to London as a journeyman printer in 1724, but the opportunity did not arise. The desire, however, demonstrates Newton's worldwide fame. Even the uneducated youth from the Colonies, long before he himself had any contact with science, knew who Newton was. His prominence at home in England was unprecedented for a scientist. He was knighted in 1705 but was not asked to travel to the royal court to receive the honor. Queen Anne came to him instead, at Trinity College, Cambridge, dined in state in the College Hall and knighted Sir Isaac in the Lodge.

To what did Newton owe this great popular fame? Surely not to the creation of the new mathematics, calculus. That might be viewed in retrospect as Newton's greatest achievement, but it could hardly be expected to have excited public adulation back in the 1690s.

Newton's great feat for his contemporaries was probably his law of gravitation and its use to solve the age-old riddle of the planets. Previous generations had devised mathematical or mechanical descriptions that could accurately plot planetary motions, but the puzzle that always remained was the apparent irreconcilability between "heaven" and "earth." Objects on earth fall to the ground because of gravity; the moon and planets and stars hang up there for ever. Mathematical laws or complex models may be able to predict the heavenly motions accurately and forecast eclipses and the strange retrogressive motions of the outer planets, but they were purely empirical. Can we escape the conclusion that the physical laws governing the heavens must be different from the physical laws on earth? This had in fact been an almost unchallengeable belief for two thousand years and had undoubtedly buttressed all sorts of superstitious notions, such as the physical existence of heaven and hell.

Newton liberated us from this dichotomy. Galileo sowed the seeds perhaps, sixty or seventy years earlier, but Newton's was the message that penetrated deep into the popular consciousness. Planetary motion was explained by him by a "celestial" law of gravitation that was the same as the "terrestial" law that governs the fall of bodies on earth. This was a scientific revolution on a grand scale, and no great intellectual effort was required to appreciate it.

Acceptance of the Newtonian world was not instantaneous, especially in France, where memories of the more metaphysical world of Descartes (a century or half-century earlier) lingered on for a while. It was, however, the French who carried out some of the critical experiments in the establishment of Newtonian laws. For example, arduous and expensive experiments were done on the orders of the French government itself to measure the curvature of the earth at different latitudes. Their results confirmed Newton's prediction (based on the law of gravitation) that the earth should be slightly flattened at the poles and should bulge out a little at the equator, in contrast to an earlier prediction by Descartes that the earth should be elongated at the poles.

The sun, the earth, and the planets are not directly relevant to the molecular events that occur when oil is poured on water, but they are only part of Newton's broader natural philosophy, in which all matter consists of particles, with forces between them that dictate the state of motion, predicting direction in space as well as speed and acceleration. Newton's now familiar laws of motion had to apply universally, to "molecules" as well as to planets. The universe around us was the testing ground for Newton's laws, but, once accepted, the laws are applicable with equal confidence to lesser phenomena on earth. "Nature is very consonant and conformable to herself," as Newton himself put it.

The point here is to establish that by 1770 there existed a consensus about physics, a belief that Newton had solved at least some of the great mysteries. This is not to imply that Newton's laws had become absolute dogma without residual opposition, but many scientists were sufficiently convinced to turn their attention to other problems: to chemistry, for example, heretofore in the hands of alchemists who wanted to grow rich by transmuting base metals into gold.*

Thus one component of Newton's legacy was a kind of release from constraint, a great broadening of the scope of the inquiry into the nature and mechanism of the world around us, some of whose basic laws we could at least tentatively assume to be known.

* It should be pointed out that substantial controversy existed regarding one aspect of Newton's physics, the matter of optical phenomena. But this was controversy about the nature of light, whether it was particulate, as Newton claimed, or a wave motion, as Christiaan Huygens in the Netherlands proposed. It was not a dispute about the existence of particles of matter and the laws that determine their motion.

3. Ultimate Particles

> Nothing exists except atoms and empty space; everything else is opinion.
>
> Democritus

The idea that matter is not infinitely divisible must surely be as old as man's thought about the world around him. Eventually one surely obtains an "ultimate particle" incapable of further subdivision or, at least, incapable of subdivision without loss of the properties that characterize the substance itself. Democritus (around 470–400 B.C.) is usually cited as the first to have stated this idea explicitly in writing, but others before him held similar conceptions of the state of matter. Democritus called his particles "atoms" (meaning "indivisible"), but the particles he had in mind probably correspond more closely to what we now call "molecules." Though too small to be visible, the particles were imagined as hard and as possessing distinct form, size, and weight. They were also thought to be in constant motion.

A discussion of this concept of atomicity inevitably raises questions. Exactly how is Democritus' atom related to our own modern view of atoms and molecules? What can we say about the corresponding Newtonian "particle"? For example, what are its mass, size, and deformability? We shall postpone these questions for the moment because, surprisingly, the very idea of "ultimate particles," so eminently sensible to us, was not universally accepted by the Greeks, or by subsequent philosophers until well into modern times. Aristotle, for example, was one who believed firmly in *continuity* of matter.

There were two reasons for this, one theological and one philosophical. In the early days the church was a vigorous opponent because Democritus' atoms were thought to be unchangeable, eternal. Only their combinations with each other led to the variety of large-scale (visible) forms. From there it was but a short (and dangerous) step to saying that coming into being and passing away are mere rearrangements of the atoms; and that you and I are temporary aggregations of atoms that will soon separate again. Democritus himself said this, and Epicurus of Samos, a famous disciple of Democritus, based an attractive theological doctrine on this notion—"gather ye rosebuds while ye may"—which ran counter to the more austere teachings of the established Christian church. As Lawrence Durrell tells it,

> Epicurus's philosophy exercised so widespread an influence that for a long time it was touch and go whether Christianity might not have to give way before it. But the poor in heart won out, don't ask me why—it is one of the great mysteries of the world.

Belief in any part of Epicurean philosophy tended to be viewed as heresy, and people were burned at the stake for it. Adherence to the Aristotelian view of the chemical world was more conducive to a long and peaceful life.

The purely scientific (philosophical) objections to fundamental particles were, however, also strong. "Ultimate particles," however logical, were invisible, entirely speculative. Moreover, the idea of a vacuum, an empty space between particles, was difficult to accept, yet this empty space is an obvious corollary. How can one claim that *all matter* consists of discrete particles without at the same time postulating a "nothing" in between? The idea is intuitively uncomfortable: for example, are there really pockets of "nothing" within a seemingly uniform body of water?

René Descartes in France had tried to cope with the problem. He conceived matter as closely packed *without any vacuum anywhere in the universe*. Because of close packing, movement of any one part of matter is only possible if other parts around it move simultaneously, so that (at the end of the chain) there is something else to take the place of the first part that is displaced. The overall result is that the universe is filled with pockets of circular motion, whirlpools, vortices. Particles, in Descartes's view, were vortices of motion.

Today it seems absurd, a tortured theory contrary to enlightened common sense. But a vacuum may have seemed just as absurd to the French philosophers, and Descartes's model remained fashionable in some quarters in France until the beginning of the eighteenth century. Voltaire, on a visit to London in 1733, is often quoted on the subject of the theory's eclipse: "A Frenchman who arrives in London, will find philosophy, like every thing else, very much chang'd there. He had left the world a plenum, and now he finds it a vacuum."*

Franklin's England, however, had become accustomed to the idea of vacuum between stars and planets in the universe, and by and large also accepted the idea of an interstitial vacuum within all matter, between "atoms." It was made easier to accept than it once had been, because it provided an attractive explanation for another phenomenon that had started to puzzle physicists—the transparency of solids when reduced to very thin plates, a topic to which Newton devoted an entire book in his *Opticks*.

Acceptance of the concept of particulate matter did not of course resolve questions about the intrinsic nature of the particles. Are atoms hard and impenetrable and truly indivisible, as Democritus viewed them? How much do they weigh, how large are they? Atomic or molecular size will be of paramount importance for the remainder of this book, and, in that connection, an extreme possibility was being discussed seriously. What if atoms, massive though they are, actually occupy very little space, or even none at all?

This idea could be held compatible with Newtonian theory, because this theory contained not only the concepts of absolute space and particles within it, but also the concept of forces between

* In fact, even in England, the idea of the "vortex atom" proved not to be completely dead. Both it and the closely packed universe, without empty space, were revived briefly in the middle of the nineteenth century. See chapter 9.

particles, acting at a distance, even when the particles were still far apart. This led Joseph Priestley, for example, to argue that force could be a *more important factor* in the universe than solid matter, which in turn led him to muse that "all the solid matter in the solar system might be contained within a nut-shell." Perhaps, extrapolating, there is really no such thing as solid matter at all?

The Jesuit scholar Ruggiero Giuseppe Boscovich (1711–1787), whose work was known in England and presumably to Priestley, had somewhat earlier carried this diminution of geometrical extension to an extreme. What if each particle is a point, having at any instant a defined position in space, possessing a certain invariant mass, but no *extension in space at all*? We need only assume the particles to be subject to forces that depend (as Newton's gravitational force does) on the distance between particles. At a relatively great distance the force is an attraction, but at close approach the ultimate force becomes a repulsion, which increases without limit as the distance diminishes. Thus, even without space-filling substance, two particles can never coincide, and the notion of impenetrability is preserved.

Wild speculation? Yes, but atomism itself was a speculation as long as the fundamental particles of matter were too small to be seen. Newton speculated that microscopes might one day become sufficiently powerful to make it possible to see them, but he was not willing to predict when that would happen or what range of size might be observed.

Where did Ben Franklin stand? In his youth he had read Newton's *Opticks*. He was aware of the colors and transparency produced in thin films, in fact mentions them in the wave-stilling account. Presumably he knew that Newton had invoked the transparency of thin solid films as support for the existence of empty space between particles. This, plus his own propensity for uncomplicated ideas, probably made him an unquestioning believer in solid ultimate particles. Franklin was aware of Descartes's ideas, but nowhere is there any suggestion that he took them seriously. His only mention of them was in 1772, in a satirical piece (satirizing much else besides Descartes), in which he asked, "What comfort can the vortices of Descartes give to a man who has whirlwinds in his bowels?"

It is not known whether Franklin was aware of the "matter in a nut-shell" idea. Priestley did not mention it in print until some years after the wave-stilling experiment. But it is not likely that the idea—solid or liquid matter made up of "atoms" that themselves occupy no space—would have appealed to Franklin. He was never the kind of man to say that things are not what they seem.

4. Distinction Between Atoms and Molecules

Newton and Franklin called them "particles," some Frenchmen called them "molecules" (meaning "little masses"), and others continued to use "atoms," as Democritus had done. But at first they all meant the same thing, the ultimate subdivision of matter for any substance, in which the characteristic properties of the substance are retained. The clear distinction that we now make between atoms and molecules did not develop until after Franklin's time and could not have been conceived by Franklin at the time of the wave-stilling experiments around 1770.

Following are some of the steps that subsequently led the way.

Clear distinction between elements and compounds. This is a distinction at the level of tangible substances, substances with common names, substances that could be seen and weighed out on a balance. Some of the substances were called "elements," defined as not being reducible to simpler substances, others were "compounds," which could be decomposed, yielding elements as products. Elements were the chemical building blocks; compounds were formed by combination of elements.

Emergence of the chemistry of gases. This first required recognition of the existence of gaseous compounds other than the familiar "air" that we breathe. The Scottish physician-chemist Joseph Black in

1754 discovered a "fixed air" (carbon dioxide) that was incapable of supporting combustion and therefore clearly not the same substance as atmospheric air. Henry Cavendish followed in 1766 with the discovery of "combustible air" (hydrogen), and finally came Priestley's many "different airs" in the 1770s, and gas chemistry was on its way. (As mentioned earlier, Priestley himself stubbornly resisted the idea that his "different airs" were actually intrinsically different substances.)

True explanation of "burning." Many people in the middle of the eighteenth century thought that something called "phlogiston" was *released* from a substance when it burned in air, and this obviously confused the definition of elements and compounds. Lavoisier swept away the cobwebs of confusion when he proclaimed oxygen to be a real substance (an element which *combined* with a combustible substance when it burned), and he published the first realistic list of the elements, shown in table 1.

Dalton's atomic theory. John Dalton, meteorologist turned chemist, is known as the "modern" father of the atomic theory, Democritus being the "ancient." He took the almost final step, in 1808, by equating elements with atoms, and guessing (or, in some cases, almost proving) that the ultimate particles of compounds are adducts of integral numbers of atoms, as indicated in table 2. Dalton did not yet use the word "molecule." He called his particles "simple" atoms, "binary" atoms, etc.

Dalton's was not quite the final step to comprehension because he assumed that the ultimate particles of elements were always single atoms. Combining weights and volumes in compound formation were not always consistent with this assumption, and the final clarification came only when it was realized that some elements in gaseous form, notably oxygen and hydrogen, consist of diatomic molecules, O_2, H_2, etc.

An interesting feature of this chemical "revolution" is that Levoisier saw in his own work a radical change in the basic fabric of the

Table 1. Lavoisier's Elements (1789)[a]

Weightless elements and gases		
Light		
Caloric (heat)		
Oxygen		
Azote (nitrogen)		
Hydrogen		
Nonmetallic acid formers		
Sulfur	Phosphorus	Carbon
plus three so-far-unknown elements that are precursors of muriatic (hydrochloric), "fluoric," and boric acids.		
Metallic elements		
Antimony	Iron	Platinum
Arsenic	Lead	Silver
Bismuth	Manganese	Tin
Cobalt	Mercury	Tungsten
Copper	Molybdenum	Zinc
Gold	Nickel	
The earths[b]		
Lime	Barytes	Silica
Magnesia	Alumina	

[a] The insight into chemistry that is obtained from the tables on this and the adjacent page was not available at the time of Franklin's experiments.
[b] The earths turned out later to be hard to reduce oxides and not true elements.

scientific method, at least as it applies to chemistry. He wrote to Ben Franklin in February 1790 (three months before Franklin died), enclosing two copies of his *Elementary Treatise*, one for Franklin personally and one for the Philadelphia Society. His letter included the following statement:

> In all chemistry treatises published since Stahl it has been customary to propose an hypothesis first and then to try an explanation of chemical phenomena on this basis. . . . I have

Table 2. Dalton's Relative Weights of Atoms (1808)

Hydrogen	1	Lime	23	Copper	56
Azote	5	Soda	28	Lead	95
Carbon	5	Potash	42	Silver	100
Oxygen	7	Strontites	46	Platina	100
Phosphorus	9	Barytes	68	Gold	140
Sulfur	13	Iron	38	Mercury	167
Magnesium	20	Zinc	56		

Examples of compound atoms

Water or steam, 1 of oxygen and 1 of hydrogen, retained in physical contact by a strong affinity and supposed to be surrounded by a common atmosphere of heat; its relative weight	8
Ammonia, composed of 1 of azote and 1 of hydrogen	6
Nitrous gas, 1 of azote and 1 of oxygen	12
Nitrous oxide, 2 of azote and 1 of oxygen	17
Nitric acid, 1 of azote and 2 of oxygen	19
Alcohol, 3 carbon and 1 hydrogen	16
plus many more.	

> sought to arrive at the truth by a succession of facts, . . . by following the torch of observation and experiment.

There is no indication in any of Franklin's writings (which, of course, were not about chemistry) that he ever followed any other philosophy. He always began with observation and experiment, and any "explanation" was solidly based on fact.

5. Heat, Light, and Electricity

Lavoisier's table of elements is really quite accurate, but the modern reader may be surprised by the inclusion of "light" and "heat" as elements. At the time that seemed quite logical. They were "imponderable fluids," apparently weightless, but sharing many other properties of particulate fluids. The similarities are especially reasonable in the case of heat. Heat is not unique in being invisible; atmospheric air shares this property and so do the newly discovered "different airs." Heat seems to flow, can be felt by the hands, and spreads outward from its source, much as air expands to fill space. Heat can be reabsorbed by a visible substance, as when ice is melted. The argument for light as an element is probably less convincing today. It was based in part on the known action of light on silver salts to produce metallic silver—photography was not far away. One could also cite Newton as a precedent. He had been a proponent of a corpuscular theory for light, as opposed to the rival wave theory of Dutch physicist Christiaan Huygens.

Given the inclusion of heat and light in this table of elements, Lavoisier's *failure to include* electricity is somewhat strange. Surely, of all the so-called "imponderable fluids," electricity is the most "tangible." In fact, we know today that it is not even weightless. Moreover, it was Benjamin Franklin whose work had created the accepted view of electricity as a fluid, he was famous for it throughout France, and he was a personal friend of Lavoisier's.

Franklin's "single fluid" theory of electricity, almost universally accepted by the time of Lavoisier, in fact stated electricity to be "particulate." In one of the letters to Peter Collinson, written July 29, 1749, he says: "The electrical matter consists of particles extremely subtile, since it can permeate common matter, even the densest metals." He goes on to state that electrical particles mutually repel one another, though they are attracted to common matter. He points out that different kinds of common matter have different degrees of attraction for electrical matter, and that electrification by rubbing reflects removal of electric particles from one substance and their appearance as an "electric cloud" on the other. The total amount of electrical fluid (number of electric particles) is absolutely conserved when this happens.

6. Science and Mathematics

To conclude this chapter, here is a section for readers without mathematical skills. Those who are terrified by an integral sign or the symbol "dy/dx". You who hide from scientists and engineers because they make you feel deficient, talking in a language that you don't understand. It may comfort you to know that Ben Franklin also knew no mathematics. Neither did Joseph Priestley, and it would not have been helpful to him (for the discovery of oxygen, for example) if he had.

Historians of science have made much of the intimate association between mathematics and physics, perhaps giving it too much emphasis. They point to Newton, who was compelled to create calculus because he could not have solved the problem of planetary motion without it. They point to the twentieth-century revolution in physics, where even the meaning of the new subjects—relativity and quantum mechanics, for example—may be virtually inexpressible by words alone. But the importance of mathematics must not be overstressed. The discovery that water is a compound of hydrogen and oxygen requires no higher mathematics. Nor does the first step toward an understanding of electricity, the realization by Franklin and others that electricity must be particulate.

And even if mathematics is used as a tool in arriving at a breakthrough in physics, as it was, for example, by Maxwell, when he derived his equations for the propagation of electromagnetic waves (around 1860), may not the final result often be capable of explanation in words alone? Michael Faraday, who was not himself much of a mathematician, certainly thought so, as evidenced by this quotation from a letter to Maxwell:

> When a mathematician engaged in investigating physical actions and results has arrived at his conclusions, may they not be expressed in common language as fully, clearly, and definitely as in mathematical formulae? If so, would it not be

a great boon to such as I to express them so?—translating them out of their hieroglyphics. . . . .

Science in fact is divided into mathematical and nonmathematical parts, with no implication that one is more important than the other. Some topics in science cannot be discussed properly without mathematics, but others do not need mathematics, and are just as accessible to the nonmathematician as to the mathematical expert. The science in this book happens to be in the latter category. It is *intrinsically nonmathematical*. It requires imaginative thinking—a mental picture of a molecule and of organized structures formed by molecules—but no calculus, differential or integral, no matrix algebra, etc. For this book, lack of mathematics is no handicap.

We must make a distinction, of course, between mathematics and arithmetic. The latter, the ability to deal with numbers, is virtually inescapable in any kind of scientific research, for all scientific research involves measurement of time, distance, mass, volume, etc. And it should be noted that Franklin had considerable agility in arithmetic, as demonstrated by his publication of some complex and ingenious "magical squares." These squares are part of what is today called "recreational arithmetic." The object is to take an $N \times N$ square, and to arrange within it the numbers from 1 to N^2, in such a way as to give the identical sum in every horizontal row and every vertical column. Franklin published squares as large as 16×16, containing the numbers from 1 to 256, with the sum of 2056 in every row and column. He also made the two diagonals add up to 2056 in this case and added other embellishments for regularity within subsections of the large square.

CHAPTER SEVEN

Franklin's Experiment: The Observation

Of the stilling of Waves by means of Oil. Extracted from sundry Letters between Benjamin Franklin, LL. D. F.R. S. William Brownrigg, M.D. F.R. S. and the Reverend Mr. Farish.

Title of the 1774 paper

1. Author's Comments

It is appropriate to divide the published account of Franklin's experiment into two parts: the observation and the interpretation. This chapter describes the observation and includes the experiment itself and the circumstances that led to it. Chapter 13 will give Franklin's attempt to interpret his observations in molecular terms. Most of the present chapter consists of a verbatim copy of Franklin's own words, with a minimum of annotation. The writing is delightful. In the words of Sir Humphry Davy, à propos of Franklin's book on electricity, "He has written equally for the uninitiated and for the philosopher; and he has rendered his details amusing as well as perspicuous, elegant as well as simple."

Franklin's paper provides convincing documentation of his scientific expertise. This was no Renaissance statesman, with a dilettante curiosity in science, but a professional with a deep comprehension of contemporary concepts. His knowledge of optics was especially good, presumably derived from reading Newton's *Opticks*. This is shown, for example, by his understanding of the refraction of light in one of the anecdotes and by the casual reference to "prismatic colors" in the account of the Clapham Common experiment itself.

The precise descriptions—characteristic of any good experimentalist—are noteworthy. For example, with reference to the disappointing practical trial in Portsmouth Harbor (described at the end of this chapter), he writes that members of the boat crew were "pouring oil continually out of a large stone-bottle, through a hole in the cork, somewhat bigger than a goose-quill."

In relation to this occasion, Franklin is conscious of the *special need for precision* in describing a failed experiment: "It may be of use to relate the circumstances even of an experiment that does not succeed, since they may give hints of amendment in future trials: it is therefore I have been thus particular." One wishes that present-day students would understand that. Their more usual procedure is to lose interest when an experiment doesn't work and to leave no record of it.

On the other hand, it must be admitted that there is a contrast between Franklin's style in the present paper and the style of his letters to Peter Collinson, written twenty-five years earlier to announce his electrical discoveries. The letters to Collinson are self-confident, commanding attention to experiment after experiment. The experiments are numbered sequentially in the longer letters, to emphasize the logic by which each previous experiment suggests the one that follows. There were no anecdotes, such as occur in the present report. There were fewer equivocal phrases—things did not "seem to be," but "were."

Age might have something to do with the contrast in style. Franklin was close to seventy when he wrote this paper, and his writing could have become more discursive with the years. It is more likely, however, that the contrast comes from the wave-stilling experiment itself. It is obviously difficult to explain. It doesn't generate new experiments, as it might have done if one had a possible explanation and wanted to test it. Franklin is less secure here, and in some places gives the impression that his mind is still searching for inspiration as he writes.*

2. Philosophical Letters

The Royal Society was founded in 1662, with a charter from King Charles II. It was originally called "The Royal Society of London for Improving Natural Knowledge." The Society began publishing its *Philosophical Transactions*, one of the world's first scientific journals, in 1665. Most of Franklin's scientific work, including the account of the wave-stilling experiments, was published here.

At first glance *Philosophical Transactions* does not seem to be a scientific journal in the modern sense at all. Look at the title page reproduced in figure 6, for example. The emphatic "INGENIOUS" in the middle of the page suggests an adventurous spirit absent from the scientific literature of today; or, more likely, probably sternly prohibited by editorial policy. When we turn to the contents, they too are different, consisting not of formal "papers," but almost entirely of letters, written from one Fellow of the Royal Society to another or from an outsider to one of the Fellows.

The subject matter of the letters is broad, ranging from precise descriptions of historic experiments (Franklin's letters on electricity to Collinson, for example) to tales of questionable veracity from travelers in foreign lands. The letters were invariably read before the Society at one of its meetings before being published. Specimens of plants or rocks or artifacts often accompanied the letters and were passed around at Society meetings and sometimes laid out for exhibition. When Collinson sent a glass tube to the Library Society

* *Note on the word "oil."* In Franklin's time the word "oil," without qualifying prefix, would have referred to oil used in the household, generally olive oil or sometimes fish oil or whale oil. Mineral oil (petroleum oil) had been known since antiquity, but saw little practical use until the nineteenth century. The distinction is important, because petroleum oil is pure hydrocarbon oil, and, as will become clear later, would not have produced the effects that Franklin observed.

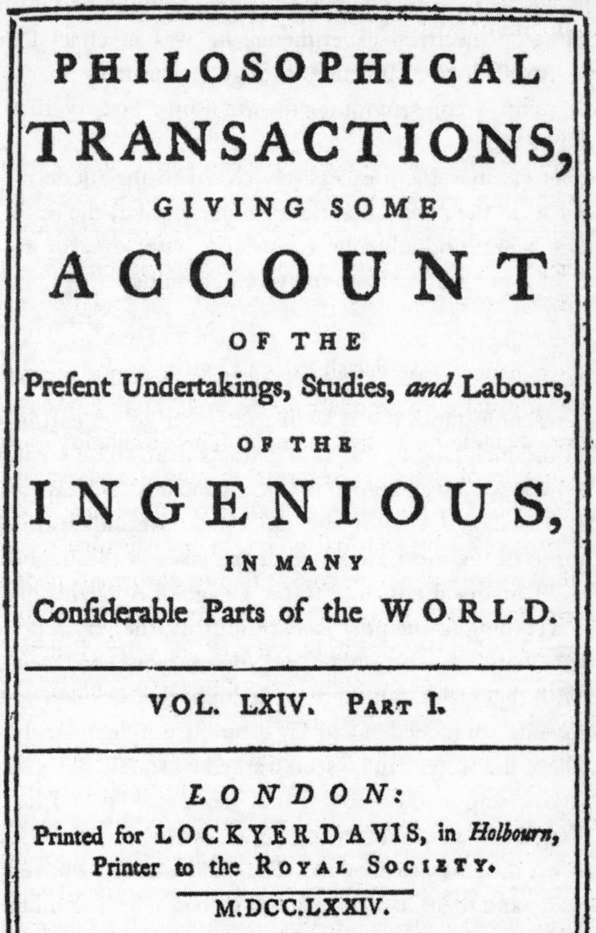

PHILOSOPHICAL
TRANSACTIONS,
GIVING SOME
ACCOUNT
OF THE
Preſent Undertakings, Studies, *and* Labours,
OF THE
INGENIOUS,
IN MANY
Conſiderable Parts of the WORLD.

VOL. LXIV. Part I.

LONDON:
Printed for LOCKYER DAVIS, in *Holbourn*,
Printer to the Royal Society.
M.DCC.LXXIV.

Figure 6. Title page of volume 64 of the *Philosophical Transactions*, where Franklin's account is published. This format had been used, with only minor modifications, beginning with volume 1 in 1665. It was replaced in 1776 by a more prosaic title page, conventional by present-day standards.

in Philadelphia in 1747, which started Ben Franklin and his friends on their own electrical experiments, he was in effect following normal Royal Society practice, though presumably many fewer curious items were sent out *from* the Royal Society than were received by them.

The letters may seem excessively casual to the modern reader, but they were the *normal* vehicle for publication at the time. Many of them were undoubtedly composed with great care, fully intended for publication from the very beginning.

3. Farish Fails a Lesson

In the present instance it was William Brownrigg who arranged for publication, including in this case extracts from three separate letters. Brownrigg, a well-known physician and amateur experimental scientist, lived at Ormathwaite, an estate overlooking Derwent Water, one of the most attractive of the lakes in the English Lake District. Ormathwaite is close to the village of Applethwaite, from which, according to the poet Robert Southey, the *very best* view of Derwent Water can be obtained. Not everyone will necessarily agree with that claim, but there is no doubt that Ormathwaite and Applethwaite are indeed in a lovely location, a mile or so above the lake, where the water can be seen framed by stately old trees, with green hills rising in the background. Ben Franklin and his friend John Pringle had visited Brownrigg at Ormathwaite in the summer of 1772, on their way to Scotland. They had walked or driven down to the lake, and there had gone out in a boat, where Franklin demonstrated the wave-stilling phenomenon for Brownrigg's benefit.

The published correspondence begins with a letter from the Reverend Mr. Farish, identified only as "a worthy clergyman at Carlisle," to Dr. Brownrigg, in which an authentic account of the demonstration was requested. Church records show that Farish was rector at St. Michael's church in Stanwix, on the outskirts of Carlisle, and, in addition, served an outlying parish as needed.

Brownrigg sent Farish's letter to Franklin in London on January 27, 1773, together with an accompanying letter of his own. Only an extract of Brownrigg's letter appears in the published text, so that we don't know whether Brownrigg, in another part of the letter, urged Franklin to write a detailed account for presentation to the Royal Society. Franklin did in fact do that in a letter from London dated November 7, 1773. "Perhaps you may not dislike to have an account of all I have heard, and learnt, and done in this way," he writes. "Take it if you please as follows." Brownrigg (who was, of course, a Fellow of the Royal Society) read extracts from all three letters to a meeting of the Society on June 2, 1774. The extracts were published in the *Philosophical Transactions* later the same year.

Farish's letter at first gives the impression of a keen and skeptical mind, but the impression soon fades:

> I some time ago met with Mr. Dun, who surprised me with an account of an experiment you had tried upon the Derwent water, in company with Sir John Pringle and Dr. Franklin. According to his representation, the water, which had been in great agitation before, was instantly calmed, upon pouring in only a very small quantity of oil, and that to so great a distance round the boat as seems a little incredible. I have since had the same accounts from others, but I suspect all of a little exaggeration.

Farish's incredulity is justified, which is why Franklin's experiment is such a good teaching aid. The result does seem exaggerated, but it is nevertheless true. We are observing a truly remarkable molecular phenomenon. Farish continues:

> Pliny mentions this property of oil as known particularly to the divers, who made use of it in his days, in order to have a more steady light at the bottom. The sailors, I have been told, have observed something of the same kind in our days, that

> the water is always remarkably smoother in the wake of a ship that hath been newly tallowed, than it is in one that is foul. Mr. Pennant also mentions an observation of the like nature made by the seal catchers in Scotland. *Brit. Zool.* Vol. IV. *Article* Seal. When these animals are devouring a very oily fish, which they always do under water, the waves above are observed to be remarkably smooth, and by this mark the fishermen know where to look for them. Old Pliny does not usually meet with all the credit I am inclined to think he deserves. I shall be glad to have an authentic account of the Keswick experiment [Keswick is the largest town on Derwent Water], and if it comes up to the representations that have been made of it, I shall not much hesitate to believe the old Gentleman in another more wonderful phaenomenon, he relates, of stilling a tempest only by throwing up a little vinegar into the air.

Poor old Farish! How revealing of character a few printed words can be! It does not seem to have occurred to him that there was a lesson to be learned from Ben Franklin. Science is not "believing," but experimenting, demonstrating. Try it for yourself! Did Mr. Farish ever think that he could have thrown a little vinegar into the air himself? Anyone who has ever visited the north of England will know that "tempests" suitable for testing are no rarity there.

Brownrigg adds a footnote to Farish's letter, giving his own example of a hearsay account of the clarification of the water that Pliny had mentioned: "Sir Gilfred Lawson, who served long in the Army at Gibraltar, assures me that the fishermen in that place are accustomed to pour a little oil on the sea, in order to still its motion, that they may be enabled to see the oysters lying at its bottom, which are there very large, and which they take up with a proper instrument."

4. Franklin's Prologue

Franklin's account begins with his first observation of the wave-stilling phenomenon in 1757. It happened on the way to England, on his first diplomatic mission as representative of the Assembly of the State of Pennsylvania. Because England was at war with France (at least in the Colonies), the transatlantic crossing was hazardous. Franklin had to wait idly in New York for several weeks before his ship was allowed to sail, initially as part of a military convoy bound for Louisburg (Nova Scotia), one of the centers of contention between France and England.

> In 1757, being at sea in a fleet of 96 sail bound against Louisbourg, I observed the wake of two of the ships to be remarkably smooth, while all the others were ruffled by the wind, which blew fresh. Being puzzled with the differing appearance, I at last pointed it out to the captain, and asked him the meaning of it? "The cooks, says he, have, I suppose, been just emptying their greasy water through the scuppers, which has greased the sides of those ships a little;" and this answer he gave me with an air of some little contempt, as to a person ignorant of what every body else knew. In my own opinion I at first slighted his solution, tho' I was not able to think of another. But recollecting what I had formerly read in PLINY, I resolved to make some experiment on the effect of oil on water, when I should have opportunity.

Franklin next refers to a related occurrence, the swinging glass lamp phenomenon, which he had noticed in 1762 on the return trip from his diplomatic mission. He uses it to introduce two hearsay accounts of his own, one from Bermuda and one from the Mediterranean, both of which report observations similar to the one he had made in 1757.

> Afterwards being again at sea in 1762, I first observed the wonderful quietness of oil on agitated water, in the swinging

> glass lamp I made to hang up in the cabin, as described in my printed papers, page 438 of the fourth edition.—This I was continually looking at and considering, as an appearance to me inexplicable. An old sea captain, then a passenger with me, thought little of it, supposing it an effect of the same kind with that of oil put on water to smooth it, which he said was a practice of the BERMUDIANS when they would strike fish, which they could not see, if the surface of the water was ruffled by the wind. This practice I had never heard of, and was obliged to him for the information; tho' I thought him mistaken as to the sameness of the experiment, the operations being different; as well as the effects. In one case, the water is smooth till the oil is put on, and then becomes agitated. In the other it is agitated before the oil is applied, and then becomes smooth.—The same gentleman told me, he had heard it was a practice with the fishermen of LISBON when about to return into the river, (if they saw before them too great a surf upon the bar, which they apprehended might fill their boats in passing) to empty a bottle or two of oil into the sea, which would suppress the breakers, and allow them to pass safely: a confirmation of this I have not since had an opportunity of obtaining. But discoursing of it with another person, who had often been in the Mediterranean, I was informed that the divers there, who, when under water in their business, need light, which the curling of the surface interrupts by the refractions of so many little waves, let a small quantity of oil now and then out of their mouths, which rising to the surface smooths it, and permits the light to come down to them.

A typical Franklinian sentence ends the paragraph:

> All these informations I at times revolved in my mind, and wondered to find no mention of them in our books of experimental philosophy.

5. Experiment at Clapham

The first paragraph is an undoctored description, and the second paragraph consists of a single sentence to tell us that this was not simply a onetime observation. The tour de force is the third paragraph, which exemplifies Franklin's exceptional scientific intuition. The spreading of the tiny volume of oil over so large an area rivets Franklin's attention (see Fig. 7). It is more interesting than the calming effect on the waves, which is what he initially set out to demonstrate.

> At length being at CLAPHAM where there is, on the common, a large pond, which I observed to be one day very rough with the wind, I fetched out a cruet of oil, and dropt a little of it on the water. I saw it spread itself with surprising swiftness upon the surface; but the effect of smoothing the waves was not produced; for I had applied it first on the leeward side of the

Figure 7. The pond at Clapham Common. Photographed by the author in 1985.

pond, where the waves were largest, and the wind drove my oil back upon the shore. I then went to the windward side, where they (the waves) began to form; and the oil, though not more than a teaspoonful, produced an instant calm over a space of several yards square, which spread amazingly, and extended itself gradually till it reached the lee side, making all that quarter of the pond, perhaps half an acre, as smooth as a looking glass.

After this, I contrived to take with me, whenever I went into the country, a little oil in the upper hollow joint of my bamboo cane, with which I might repeat the experiment as opportunity should offer; and I found it constantly to succeed.

In these experiments, one circumstance struck me with particular surprise. This was the sudden, wide and forcible spreading of a drop of oil on the face of the water, which I do not know that anybody has hitherto considered. If a drop of oil is put on a polished marble table, or on a looking glass that lies horizontally; the drop remains in place, spreading very little. But when put on water it spreads instantly many feet round, becoming so thin as to produce the prismatic colors, for a considerable space, and beyond them so much thinner as to be invisible, except in its effect of smoothing the waves at a much greater distance.*

His thoughts now begin to turn (but only for a few lines) to interpretation:

It seems as if a mutual repulsion between its particles took place as soon as it touched the water, and a repulsion so

* The bamboo cane with an "upper hollow joint" merits a comment. C. H. Giles, in an article written for a chemical journal in 1969, has conjectured that it may have been a physician's cane, with a compartment for holding small quantities of drugs.

> strong as to act on other bodies swimming on the surface, as straws, leaves, chips, &c. forcing them to recede every way from the drop, as from a center, leaving a large clear space. The quantity of this force, and the distance to which it will operate, I have not yet ascertained;

And again at the end pure Franklin—

> but I think it a curious enquiry, and I wish to understand whence it arises.

6. Subsequent Observations

Now a new observation is introduced, derived from a repetition of the spreading experiment for the benefit of "the celebrated Mr. Smeaton" near Leeds, during Franklin's trip north to visit Brownrigg in the Lake District.* Dead flies, drowned in oil, appeared to come back to life when thrown upon the water. Franklin (presumably without having had the opportunity to deliberate on the matter at length) quickly disposed of the possibility of resuscitation by showing that oil-soaked chips or paper tossed on the water moved as vigorously as the dead flies. An interesting aspect of this new observation in the historical context of this book is that when Rayleigh repeated Franklin's experiment more than one hundred years later (see chapter 10), he used the vigorous

*The celebrated Mr. Smeaton was John Smeaton, indeed a man of distinction, a member of the Royal Society, and an occasional visitor to the Lunar Club. He began as a maker of scientific instruments, but soon turned to the more profitable area of largescale structural engineering. He acquired great renown for the rebuilding of the Eddystone lighthouse, England's famous maritime beacon at the entrance to Plymouth Harbor. He introduced the term "civil engineering" into the language, to distinguish people such as himself from military engineers who at the time were being trained at the Royal Military Academy at Woolwich.

Figure 8. Derwent Water, as seen from a point close to Ormathwaite. Dr. Brownrigg and his guests would have had to walk or ride a mile or two to get to the edge of the lake. Photographed by the author in 1985.

motion of chips of camphor on the surface as a measure of the part of the surface that was not covered by oil.

> In our journey to the north, when we had the pleasure of seeing you at Ormathwaite, we visited the celebrated Mr. SMEATON near Leeds. Being about to shew him the smoothing experiment on a little pond near his house, an ingenious pupil of his, Mr. Jessop, then present, told us of an odd appearance on that pond, which had lately occurred to him. He was about to clean a little cup in which he kept oil, and he threw upon the water some flies that had been drowned in the oil. These flies presently began to move, and turned round on the water very rapidly, as if they were vigorously alive, though on exam-

ination he found they were not so. I immediately concluded that the motion was occasioned by the power of repulsion above mentioned, and that the oil issuing gradually from the spungy body of the fly continued the motion. He found some more flies drowned in oil, with which the experiment was repeated before us. To shew that it was not any effect of life recovered by the flies, I imitated it by little bits of oiled chips and paper cut in the form of a comma, of the size of a common fly; when the stream of repelling particles issuing from the point, made the comma turn round the contrary way. This is not a chamber experiment; for it cannot well be repeated in a bowl or dish of water on a table. A considerable surface of water is necessary to give room for the expansion of a small quantity of oil. In a dish of water, if the smallest drop of oil be let fall in the middle, the whole surface is presently covered with a thin greasy film proceeding from the drop; but as soon as that film has reached the sides of the dish, no more will issue from the drop, but it remains in the form of oil, the sides of the dish putting a stop to its dissipation by prohibiting the farther expansion of the film.

What Franklin says here is of course not correct. If the "smallest drop" is made small enough, then the dish on the table becomes equivalent to the pond in the open. See the experimental arrangement of Devaux in figure 1 in the Introduction.

This part of the paper concludes the way it started, with two more anecdotes, followed by a short but purposeful final sentence.

Our friend Sir JOHN PRINGLE being soon after in Scotland, learnt there, that those employed in the herring fishery, could at a distance see where the shoals of herrings were, by the smoothness of the water over them, which might possibly be occasioned, he thought, by some oiliness proceeding from their bodies.

A gentleman from Rhode-island told me, it had been remarked that the harbour of Newport was ever smooth while any whaling vessels were in it; which probably arose from hence, that the blubber which they sometimes bring loose in the hold, or the leakage of their barrels, might afford some oil, to mix with that water, which from time to time they pump out to keep the vessel free, and that same oil might spread over the surface of the water in the harbour, and prevent the forming of any waves.

This prevention I would thus endeavor to explain.

The text that follows this line in the paper makes it clear that what Franklin meant here by "explain" and earlier by "I wish to understand whence it arises" is an explanation on the basis of molecular properties. We shall, however, postpone this interpretive part of Franklin's paper until chapter 13.

7. Trial at Sea

A little later (presumably some time in 1773), Franklin addressed himself to the possibility that practical benefits might derive from the wave-stilling effect. His question was whether oil on water could dampen not only little waves on a pond, but also the high waves of an ocean swell, waves that can break over ships and do much damage.

Franklin was demonstrating the wave-stilling experiment on the pond in London's Green Park to Count Bentinck of Holland, his son Captain Bentinck, and "the learned Professor Allemand." Count Bentinck was the son of William Bentinck (the first Earl of Portland), who had come to England in 1689 in the retinue of William of Orange, when the latter became co-monarch of England—part of the royal partnership of "William and Mary." Captain Bentinck was a career naval man, since 1770 commander of the *Centaur*, a guard ship stationed at Portsmouth. According to the

British *Dictionary of National Biography*, he possessed great ingenuity in mechanical pursuits and effected many useful nautical improvements. The "learned professor" was probably another Dutchman, Johannes Allemand, who was professor of philosophy and natural history at the University of Leiden.

Count Bentinck had had a letter from Batavia (in the Dutch East Indies) in which it was claimed that oil poured into the sea had saved a Dutch ship during a storm. Franklin asked for and received a copy. The letter was written in French, and Franklin gives both the original and an English translation in the 1774 paper. (The translation was presumably made by Franklin himself—I could find only one possible error of translation.) Discussion of the letter and the story told in it led Franklin to express his own thoughts on the matter of practical applications:

> I mentioned to Captain BENTINCK, a thought which had occurred to me in reading the voyages of our late circumnavigators, particularly where accounts are given of pleasant and fertile islands where they much desired to land upon, when sickness made it more necessary, but could not effect a landing through a violent surff breaking on the shore, which rendered it impractical. My idea was, that possibly by sailing to and fro at some distance from such lee shore, continually pouring oil into the sea, the waves might be so much depressed and lessened before they reached the shore as to abate the height and violence of the surff, and permit a landing; which, in such circumstances, was a point of sufficient importance to justify the expense of the oil that might be requisite for the purpose.

Captain Bentinck, making use of the resources at his command, promptly invited Franklin to Portsmouth, provided the boats required for an experimental test, and accompanied Franklin on the occasion.

Accordingly, about the middle of October last, I went with some friends to PORTSMOUTH; and a day of wind happening, which made a lee-shore between HASLAR HOSPITAL and the Point near JILLKECKER, we went from the Centaur with the long-boat and barge towards that shore. Our disposition was this: the long-boat was anchored about a quarter of a mile from the shore; part of the company were landed behind the Point (a place more sheltered from the sea) who came round and placed themselves opposite to the long-boat, where they might observe the surff, and note if any change occurred in it, upon using the oil. Another party, in the barge, played to windward of the long-boat, as far from her as she was from the shore, making trips of about half a mile each, pouring oil continually out of a large stone-bottle, through a hole in the cork, somewhat bigger than a goose-quill. The experiment had not, in the main point, the success we wished, for no material difference was observed in the height or force of the surff upon the shore; but those who were in the long-boat could observe a tract of smoothed water, the whole length of the distance in which the barge poured the oil, and gradually spreading in breadth towards the long-boat. I call it smoothed, not that it was laid level; but because, though the swell continued, its surface was not roughened by the wrinkles, or smaller waves, beforementioned; and none, or very few white-caps (or waves whose tops turn over in the foam) appeared in that whole space, though to windward and leeward of it there was plenty; and a wherry, that came round the point under sail, in her way to Portsmouth, seemed to turn into that tract of choice, and to use it from end to end, as a piece of turnpike-road.

It may be of use to relate the circumstances even of an experiment that does not succeed, since they may give hints of amendment in future trials: it is therefore I have been thus particular.

In the light of the negative result in this large-scale experiment, what do we make of stories claiming greater success? The tale told by the Venerable Bede, for example, at the beginning of this book, about the oil provided by Bishop Aidan to assure a smooth passage for Princess Eanflaed's voyage from Kent to Northumberland. Should we now give some credence to the Venerable Bede himself, who ascribed the action of the oil not to natural law but to divine intervention? "He who judges the heart showed by signs and miracles what Aidan's merits were," said Bede. And he relates three "miracles," of which the calming of the tempestuous sea for Princess Eanflaed is only one.

According to modern scientific judgment, natural law should suffice. Some beneficial wave-stilling (as distinct from "spreading" alone) is obtained even on the open sea if the oil can be continuously replenished. The swells cannot be prevented, but the formation of small surface ripples that lead to the breaking of the larger waves should be diminished. For a relatively small boat, that ought to significantly improve the chances of riding out a storm. (See chapter 9 for later tests at sea.)

CHAPTER EIGHT

How Small is a Molecule? The Calculation Franklin Did Not Make

1. A Layer One Molecule Thick

HERE'S a puzzle, posed before but never answered. Ben Franklin knew that a drop of oil, when spread out over an area of half an acre, becomes astonishingly thin, thin enough to exhibit the prismatic colors, and even thinner than that, so that the presence of the oil would have been undetectable but for the smoothing of the waves. Why did he not calculate precisely how thin? He was a practical man, accustomed to weights and measures, and he must have known the formula

$$\text{area} \times \text{thickness} = \text{volume}$$

If into this formula we enter the volume of a teaspoon (2 *cc*) and the area of half an acre (2,000 m^2), we arrive at the incredibly low value of 10^{-7} *cm* for the film thickness—less than one ten millionths of an inch is what Franklin would have said. Why didn't he say it?

Franklin's lack of mathematical background cannot be used as explanation, for this involves only arithmetic and Franklin had good facility with numbers. Franklin had in fact made a similar sort of calculation at least once before in his life and tells about it in his *Autobiography* (written, you will recall, in his old age). The occasion was the visit to Philadelphia of an itinerant preacher from Ireland, the Reverend Mr. Whitefield, an evangelist, follower of John Wesley in promoting the tenets of Methodism. Mr. Whitefield had a loud and clear voice. He preached one evening from the top of the courthouse steps on Market Street. Franklin, who was far back among the hearers, became curious to learn how far the preacher's voice would carry, so he moved back from the edge of the crowd and found the preacher's voice distinct as far as Front Street, a distance of two hundred feet. He writes as follows:

> Imagining then a semi-circle, of which my distance should be the radius, and that it were fill'd with auditors, to each of whom I allow'd two square feet, I computed that he might well be heard by more than thirty thousand.

This calculation removed some of the skepticism with which Franklin had until then read stories that told of "generals haranguing whole armies."

So why not here imagine the drop of oil, extending a centimeter or so above the water (or a quarter of an inch, or the thickness of a finger, or whatever unit you choose)? And imagine (as in Fig. 9) how much thinner it must become as it spreads out to a large area around its original position?

If Franklin had made the calculation, had arrived at the figure of about 10^{-7} cm for the oil layer thickness, would he have realized that 10^{-7} cm must be a measure of molecular dimensions? There is no doubt at all that he thought in particulate terms when he analyzed the *forces* that might be involved in the spreading phenomenon. Would he not have used the same concepts when thinking about *geometrical dimensions*? Figure 10 poses the problem in terms of ultimate particles (molecules, atoms), initially heaped in a mound, and then tumbling off the mound, spreading all around, until there are no more to tumble. The simplest possible conclusion would seem to be that the point at which there are no more particles to tumble is the point at which the layer of oil on the water has been

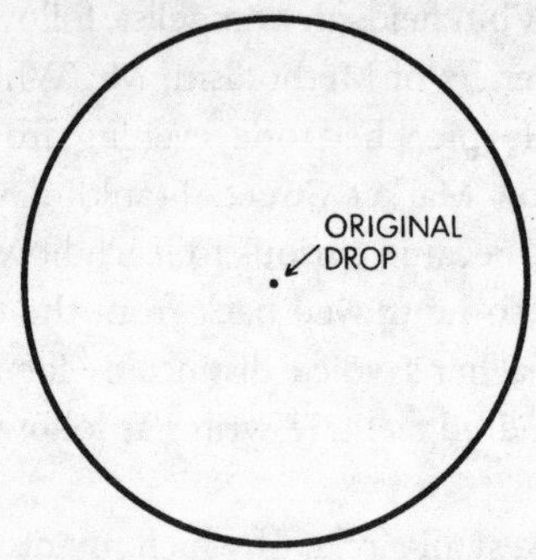

Figure 9. Visualization of the spreading of the oil. Imagine an approximately hemispherical drop that just covers the small black circle. Imagine it spreading over the entire area of the large circle. This would represent an increase in area by a factor of 5000. What Franklin observed, a teaspoonful going to half an acre (a factor of 10^7), is a further 2000-fold increase in area beyond what we see here! The area and thickness calculated from Franklin's result are now known to be correct for olive oil to well within the experimental error of estimating acreage by eye.

[Lord Rayleigh in 1890 also comments on the difficulty of comprehending very large numbers, and the need for analogies to aid visualization. He points out that the ratio of the oil film thickness to one inch is about the same as the ratio of a second of time to an entire year.]

reduced to a layer of single molecules. And that 10^{-7} cm must then be the "thickness" of an individual molecule.

Had Franklin actually reached this conclusion, then his paper would have become one of the classics of scientific literature, repeatedly reprinted like his papers on electricity. As I said in an earlier chapter, most scientists at the time, including Franklin, accepted the concept of a "molecule" as the smallest particle that retained the attributes of any chemical substance. Newton even stated his belief that these ultimate particles might some day be seen by the eye through a microscope. But no numerical estimates of molecular size existed in 1770, no one had any idea how small

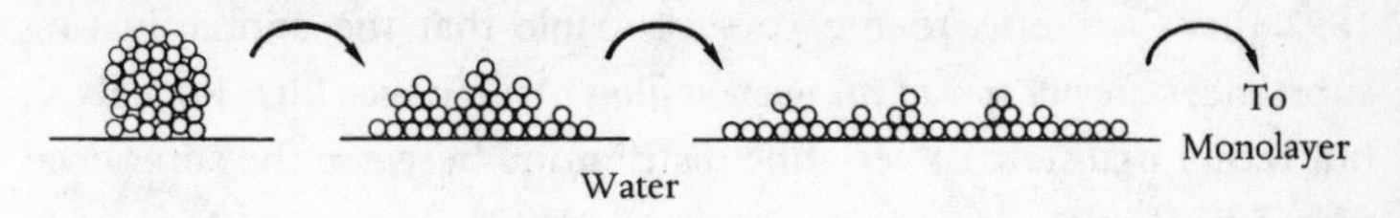

Figure 10. Another visualization. Think of the oil drop as a mound of molecular oil particles, on a huge body of water particles. Imagine the oil particles tumbling from the mound, later tumblers pushing aside the already tumbled. When will the process stop? Why should it stop at all unless there are no more particles to tumble—when every single oil molecule rests on the water?

the ultimate particles might be. Father Boscovich and Franklin's friend Priestley even entertained the notion that *molecules might be points, with no extension in space at all.*

Franklin had actually correctly determined the scale of magnitude of molecular dimensions, the first person ever to do so, but he did not recognize it. Moreover, many others had knowledge of Franklin's result or had actually seen his demonstration, including Joseph Priestley. They also failed to recognize the significance of Franklin's result. They did not even, to anyone's knowledge, make the simple calculation of film thickness based on the spreading area. The question "Why not?" applies to them as much as to Franklin.

This kind of intellectual-historical puzzle is a fit topic for a Ph.D. dissertation but in this book must remain unsolved. It is possible that notions of repulsive forces obstructed clear thinking. "It seems as if a mutual repulsion between its particles took place," Franklin said of the spreading oil. But that should not have prevented the calculation of the ultimate "area," only the interpretation thereof.

It is interesting that caution about the interpretation still existed 120 years later, when Franklin's experiment was repeated (see chapter 10). "The thickness of the film of olive-oil, calculated *as if continuous* [italics mine] . . . is about 2×10^{-7} cm," Lord Rayleigh said in 1899. Agnes Pockels was even more skeptical, writing in

1892 that "it seems to me very probable that the contaminating substances *are not spread in coherent films at all* [italics Miss Pockels's], but rather in a state of very fine distribution between the superficial water molecules. . . ."

Even with doubts about the cohesiveness of the oil film, it would seem to have been appropriate to make the calculation. Assuming that no water molecules intervene between oil molecules on the surface, then the oil molecules must extend about 10^{-7} cm above the surface. If there are intervening water molecules, then the height of individual oil molecules would need to be more than 10^{-7} cm. In no way could the molecular height be *less than* 10^{-7} cm. If the calculations had been made, where would it have left Priestley's notion that molecules might have no extension in space at all?

2. Another Puzzle?

Here's another question about the way our minds work. Why did Franklin's experiment attract so little attention in later years? Why is even Lord Rayleigh's late nineteenth-century repetition largely unremembered? It is not even mentioned in R. B. Lindsay's otherwise scholarly biography of Rayleigh in the 1977 *Dictionary of Scientific Biography.*

In this case one can make an educated guess—that the neglect of these experiments is related to what one might call the "instrumental revolution" in science. By 1770 we had entered the age of the telescope and microscope, by Rayleigh's time we had the spectroscope, and shortly thereafter the cathode-ray tube and X rays. Scientists had begun their love affair with complex apparatus. They became obsessed with the dogma that scientific progress depends chiefly on new technical advances in measuring ability and not (or at least not primarily) on what goes on in the scientist's brain. We still today turn up our noses at people who do anything as ridiculously simple as measuring the area to which a drop of oil expands on water. For some modern scientists, priding themselves on their ability to obtain grants for expensive and impressive apparatus, such people may even be considered a threat.

The poet Rupert Brooke, in 1914, in a list of things he loved, included "the keen unpassioned beauty of a great machine." Brooke died in 1915, from the machinery of war, but he had spoken for many who survived.

3. Numbers Tell the Tale

"Shout out your numbers, loud and strong"

song of the U.S. Field Artillery

Pythagoreans believed numbers to be divine and virtually worshiped them. Numbers were the elements of all things. The heavens were viewed as a musical and numerical scale. The number "10" was perfect, and they believed that it was the number of "moving" heavenly bodies (planets, etc.). And they held many more mystical numerical notions.

I myself have always liked numbers, thought of them as friends. When I started my scientific career, which was before the advent of computers, I derived genuine visceral pleasure from filling pages of "data forms" with columns and rows of numbers. Sometimes they were numbers derived from an experiment, from which other numbers had to be calculated, and so forth in progression. At other times the numbers in the columns were building blocks for some theoretical calculation: x in column 1, x^2 in column 2, x^3 in column 3, then $1 + x + 2x^2 + 3x^3$, etc., each laboriously computed from the column before, until the desired function emerged triumphantly in the final column.

Needless to say, I have always been fascinated by those particular milestones in scientific history in which a number by itself is decisive. *Numerical data*, of course, play a role in virtually all the sciences. Even intrinsically nonnumerical subjects such as taxonomy depend on numerical observations, such as the length of the bill as a factor in bird classification. In the same vein, the discovery of oxygen by Priestley and Lavoisier depended on numerical measurement of gas volumes and the weight of the oxide from which it was derived. But that is different from what I am describing here, the special examples where the number itself is the "result," occupies the center of the stage, is fixed in the imagination and pondered over.

Franklin's experiment is one of these. The expansion in the area of contact between the oil and its supporting surface by a factor of 10^7 is staggering. And it leads, as we have seen, to the estimate of molecular dimensions as being of the order of 10^{-7} cm, an equally astonishing result, given that we cannot see anything that small with our unaided eyes or even with a microscope. (The limit of resolution of a light microscope is only about 10^{-4} cm.)

An even more impressive number can be estimated from the molecular dimension. How many molecules of olive oil make a teaspoonful? Suppose, for example, that each molecule is a sphere with a diameter of 10^{-7} cm. The volume per molecule would then be 5×10^{-22} cm^3, and the number of molecules in the volume of a teaspoon would be about 3×10^{21}, even after allowance is made for the fact that spheres cannot be packed together into a bulk fluid without leaving empty space between the spheres. 10^{21} is one thousand million million million. Molecules are spectacularly small!

A different example, from the realm of biology, is provided by William Harvey's "discovery" in about 1615 of the circulation of blood. It also hinges on a very large number that can be obtained by simple arithmetic from relatively routine measurements. Harvey noted that the pump action of the heart pushes blood forcibly into the arteries, away from the heart, to all parts of the body. He measured the amount per heartbeat—about two ounces. As everybody knows, a normal heart beats about seventy-two times per minute. Multiplying seventy-two by two tells us that 144 ounces of blood must be ejected from the heart per minute. That is the same as nine pounds per minute, or 540 pounds per hour. It was commonly believed at the time that blood might be derived from the air we breathe or possibly from food, but it is out of the question that we can get 540 pounds of anything from those sources—not in a day, not in a week, much less in an hour. The only possible explanation is that the blood must be *circulating* in the body, returning to the heart over and over again to be pumped anew into the arteries.

Spectacularly large numbers—distance of the stars, age of the earth, number of molecules in a dish—of course fascinate us the most, but more pedestrian numbers can often be just as effective scientifically. In chapter 18 we shall cite a surface area measurement that involves only "a factor of two," but nevertheless has dramatic impact on our understanding of biology. Even much smaller numbers can be critically important. A discrepancy in the measured density of a sample of nitrogen of one part per 1000 led to the discovery of argon (see chapter 10). And a discrepancy of forty-three seconds per century in the orbit of the planet Mercury was one of the factors in the genesis of Einstein's theory of relativity.

CHAPTER NINE

One Hundred Years Later. Science Comes of Age

1. Heart of the Empire

THE story moves to London in 1890, the heart of the British Empire, the most important city in the world.

Queen Victoria was on the throne, nearing the end of her long reign. She was Empress of India now (since 1876), as well as monarch of the British Isles, and nominal ruler over the vast areas of Africa and Asia that had been annexed to the empire to provide new markets and to safeguard trade routes. At home it was the era of Liberal/Conservative alternation in Parliament, where members battled over such issues as Home Rule for Ireland. As Gilbert and Sullivan put it:

> I often think it's comical
> How Nature always does contrive
> That every boy and every gal,
> That's born into the world alive,
> Is either a little Liberal,
> Or else a little Conservative!

Lord Rayleigh, who took up the study of oil on water where Franklin had left off, was born a strict Conservative, with a heritage of privilege from the day of his birth, and he uncompromisingly remained a political Conservative throughout his life. His heritage profoundly affected his scientific life. In fact, his personal circumstances (such as ill health as a youth) would probably have prevented him from even thinking of a scientific career had it not been for his privileged position.

The century since Benjamin Franklin's London sojourn had seen enormous change, especially in travel and communications as well as in other aspects of daily life that would have most affected and astonished him. The many practical uses for electricity, for example. How would Franklin have reacted to them? He could legitimately consider himself to be one of the *discoverers* of electricity, but at the time of the discovery, around 1750, he could see in it "nothing of use to mankind." Now the streets were beginning to be lit with electricity and electric motors were coming into common use.

There were 250,000 miles of railroad track worldwide in 1890, including in London the first metropolitan underground railroad, which had come into service in 1863. Great steamships, powered by 30,000-horsepower engines, were crossing the Atlantic Ocean in seven or eight days. News traveled even faster, virtually instantaneously by telegraph, even between England and America now linked by transatlantic cable.

Advances in science were an integral part of this era of technological progress. As a matter of fact, Lord Kelvin's *theoretical physics* had a direct role in making the transatlantic cable possible and (he was William Thomson at the time) this earned him a knighthood in 1866. More generally, the century had witnessed a momentous era in physics, with the establishment of the foundations of thermodynamics (Joule, Helmholtz, Kelvin, Gibbs) and of the electromagnetic theory (Faraday, Maxwell).

There had been equally explosive growth in chemistry. The catalog of compounds now ran to tens of thousands, organic as well as inorganic, natural as well as synthetic. There were formulas for them as well, not just a count of how many of each kind of atom in a molecule, but a representation of how the atoms were joined to each other. Molecular properties were known to depend on this—not just on "composition," but also on the specific arrangement of atoms within a molecule and on the presence of particular

Ammonium cyanate

Urea

Ethyl alcohol

Dimethyl ether

Figure 11. Chemical structural formulas for two pairs of isomeric compounds which contain the same atoms differently arranged. Single lines represent ordinary chemical bonds joining one atom to another. The bonds are like drive shafts, in the sense that groups of atoms joined by the bonds can rotate freely about them. Double lines represent "double bonds"—they are like two axles side by side, with no rotation possible. The "triple bond" in the cyanate ion is of course also rigid.

Ammonium cyanate is an ionic compound, the NH_4^+ and OCN^- portions being held together in the crystalline state by electrostatic attraction. Wöhler would not have known this when he converted ammonium cyanate into urea, and he would have made no distinction between different kinds of chemical bonds.

groupings of atoms ("radicals") that can tenaciously retain their chemical characteristics even when built into otherwise quite dissimilar molecules. Figure 11 shows the principle of the "radical" idea for those readers without previous contact with chemistry. It gives the extended formulas for two pairs of "isomeric" molecules, meaning molecules of the same atomic composition but which differ in arrangement. One pair shows urea, the first "organic" molecule ever synthesized in a laboratory (in 1828, by Friedrich Wöhler), and ammonium cyanate, the inorganic substance from which it was made. The other pair shows two purely organic isomers, ethyl alcohol and dimethyl ether. The groupings "CH_3," and "OH" are examples of "radicals." The presence of the latter on the alcohol molecule gives it quite different properties from those of the ether.

Last, but not least, there were dramatic new revelations in biology and anthropology. The publication of Darwin's *Origin of Species* in 1859 and the discovery of Neanderthal man in Germany in 1856 may have had no direct effect on urban life, but their long-range influence on society was considerable—man's conception of himself and his place in the universe had been irrevocably shaken, with repercussions that we still feel today.

In England the new view of progressive evolution by natural selection had a curious side effect. Darwinism, in the popular mind, was readily extended to races and societies, suggesting that to be "on top" is the natural result of being "fittest," and, England being clearly on top, the conclusion was obvious. Gilbert and Sullivan again say it well:

> For he might have been a Roosian,
> A French or Turk or Proosian,

Ethyl alcohol and ether are shown at the bottom as constructed from three-dimensional "atomic models," which reflect the correct size of the atoms and the correct directions for bonds to neighboring atoms. See caption to fig. 13 at the end of this chapter.

Or perhaps Ital-ian,
But in spite of all temptations
To belong to other nations,
He remains an Englishman.

The song was sung, no doubt, to loud applause all around.

Meanwhile, on the Continent, "Proosians" had become "Germans," as a result of Count Bismarck's statesmanship and Prussia's victory over France in the Franco-Prussian War of 1870–1871. William I, King of Prussia, had become the first German Kaiser at the end of that war, an emperor to rival the British Queen in splendor and power. This shift in political power is reflected in scientific activities of the period. Germany had become the dominant country on the European continent even in pure science. Discussing Benjamin Franklin earlier, it was logical for us to talk of a "French Connection." For English scientists around 1890 it was more natural to turn to Germany for interaction and even for inspiration, as we shall see in the following chapters. (So it was, at least, for physicists. In biology the change was probably less evident, for Louis Pasteur of France was the dominant figure.)

America, now a century old, was thriving—Franklin would have been proud—but its science was still far behind any of the major European nations. A. Whitney Griswold, former President of Yale University, has written about this period: "Science . . . was in its infancy, its followers little appreciated, their equipment primitive, and their labors, like their thoughts, concerned more with practical than with theoretical matters."

2. Science as Profession

European physical and chemical science around 1900 has a "modern" look. The names are familiar, not as belonging to shadowy figures from the history of science, but in the sense that what these scientists did is textbook material, part of our own era. They represent the early part of our era, to be sure, but their work is taught in *today's* courses as part of the currently required subject matter. Their papers were published in familiar journals, as prominent today as they were back then. They are still cited as originators of methods and equations that we actually use, and citations are given by volume and page number in the same format as for more recent work. Furthermore some received the same token of world recognition that we use today, the Nobel Prize, first awarded in 1901.

It would be wrong, though, to think of the lives of the scientists of the period from the perspective of our own day. The last two decades of the nineteenth century are in fact a fascinating midpoint between the mostly amateur science of Franklin's time (at least in Britain and America) and the highly organized scientific professions of today. The intellectual format of research may have been much as it is today, but the personal aspects of a scientific career were often still as they had been in the times of Ben Franklin and his friends. There was still a sense of adventure and the freedom to pursue chosen goals, and both these conditions stimulated freedom to think independent thoughts. One reason was perhaps that science (in Britain and in the United States) still offered uncertain financial rewards, and those who chose science for a career often had independent incomes to pay for their independent thoughts. But even those for whom this reason did not apply shared in the spirit of freedom.

In particular, scientists at the end of the nineteenth century seem to have been able to devote some of their energies to thinking about a variety of fundamental problems outside their immediate research goals. If such thinking proved productive, generated new ideas, it could interrupt previous research and produce significant work in an unexpected direction, and there was rarely an actual or implied obligation not to change course in this way. Rayleigh's and Roentgen's experiments on the spreading of oil on water, which will be described below, are good examples. But they are not isolated examples. Rayleigh's work leading to the discovery of argon and Roentgen's work leading to the discovery of X rays (both projects that resulted in the award of Nobel Prizes) were both equally unpredictable departures from previously established research directions.

It is quite different today, when science has become a trade, a job that will earn an assured living, where the need to preserve status and income inevitably help shape a scientific career. One way or another, most of today's scientists, once beyond their student days, have little freedom to choose specific research problems to work on. The choice may be dictated to them directly by the management of an industry or indirectly by competition for research grants. In either case, spontaneous changes in direction are virtually forbidden. Even a modest research grant would constitute an obligation to pursue the goals set forth when applying for the grant. And an application for a new research grant in a field where the applicant has no preexisting demonstrable expertise would almost certainly be denied.

An event of special importance for the life of Lord Rayleigh—and indeed for the future development of all of physics—was the founding of the Cavendish Laboratory at Cambridge University in 1871 and the simultaneous creation of the Cavendish Professorship for Experimental Physics. Until about that time hands-on experimental work (as distinct from "demonstrations") was not part of the course of instruction at any university, but teaching laboratories had now been installed at Oxford and London, and Cambridge desired to follow suit. The necessary money was provided by William Cavendish, the 7th Duke of Devonshire, who was then the Chancellor of the University, and the laboratory and professorship were named after him.*

The manner of selection and appointment of the first Cavendish Professor is worth a note, because it is another example of the gulf

*Henry Cavendish, the famous chemist/physicist of Franklin's day, mentioned in chapter 6, had been a member of the same family, a nephew of the 3rd Duke of Devonshire. He had left much of his work unpublished, in sealed packages which were in the possession of the 7th Duke. These packages were opened only after the Cavendish Laboratory was created. Their contents were edited and published by Cambridge University Press in 1879.

between the late nineteenth century and the mid or late twentieth century. The university's first choice for the position had been William Thomson (later Lord Kelvin), who was professor at the University of Glasgow and who had actually created an embryonic laboratory for students in the college basement, formerly occupied by a wine cellar and a coal cellar. It was not a teaching laboratory in the formal sense of the intended Cavendish Laboratory, but it was a place where Thomson could use students to help him with his own research, and where students could thereby gain experience that was not at that time to be had anywhere else in England or Scotland.

Thomson declined the appointment, but he volunteered his services in inviting the next choice, who was Hermann Helmholtz, then professor of physiology at the University of Heidelberg and a personal friend of Thomson's. On January 28, 1871, he sent a handwritten letter to Helmholtz:

My dear Helmholtz

I have been asked by Stokes, and by the Master and Tutor of my College at Cambridge, to write to you asking if you could be induced to accept a new professorship of Experimental Physics to be established there. It is much desired to create in Cambridge a school of experimental science, not merely by a system of lectures with experimental illustrations but by a physical laboratory in which students under the direction of the professor and his assistant or assistants, would perform experiments . . .

The letter gave details about the salary that could be anticipated and pointed out that the duties of the professorship would occupy only twenty weeks out of the year, and that the laboratory would be available to Helmholtz for his own research the rest of the time. The letter was written from London, where Thomson was attending a meeting of a committee to "advise the Admiralty on Designs for Ships of War," and Helmholtz was instructed to address his answer to Thomson's Glasgow address.

Helmholtz felt unable to accept the invitation because he had already accepted a chair of physics at the University of Berlin, but his answering letter to Thomson has not been preserved. The negative decision, however, is known to have been transmitted before February 9, because Helmholtz's wife reported it on that day to her mother, with the indication that she herself would rather have lived in Cambridge than in Berlin. By February 13 the university must have been informed and its appointments committee must have met to decide on their next choice, for the committee wrote to James Clerk Maxwell on that day (to his estate in Glenlair, Scotland), and he promptly accepted. The chronicler of this exchange (N. Kurti of the University of Oxford) comments on how expeditiously this matter was dealt with, and also notes that "the postal services were gratifyingly swift."

Another noteworthy feature of the time (and another contrast with today) is the surpassing brilliance of the *Encyclopaedia Britannica*. It set out to be the ultimate reference in all fields, but made special efforts in the sciences, inviting the most prominent scientists to write authoritative reviews. They were addressed to a broad audience as far as that could be done, but no compromise with scholarly exactness was asked or expected. Mathematics, in particular, was never avoided as a deference to the general reader. The wave theory of light, for example, cannot be discussed without complex mathematics, and the article on that subject (written for the ninth edition by Lord Rayleigh) accordingly has a highly mathematical content.

The ninth edition, published serially between 1875 and 1889, and the eleventh edition, begun in 1903 but published as a unit between 1910 and 1911, are the two most memorable editions for scientific articles. They can be found in most good libraries today, and browsing through the volumes gives today's reader a marvelous opportunity to capture the "state of the art" of all sciences at this one brief period of time. They also illustrate the versatility of the great scientists of the period. The ninth edition, for example, has ten articles written by James Clerk Maxwell on such topics as atoms, capillary action, diffusion, Faraday, and "Physical Sciences" in general; Lord Rayleigh contributed articles on optics and on the wave theory of light. In the eleventh edition Rayleigh wrote the article on capillary action, but explains that it is only a revision of Maxwell's earlier article, and he retained the authorship initials "J. C. M." at the end of the article in addition to his own. Rayleigh wrote four other reviews for that edition, on argon, diffraction, interference, and the sky.

3. Molecular Dimensions

What was the state of knowledge about molecules now, more than one hundred years after Ben Franklin? Strangely, there was a dichotomy, not at all what is expected from the popular notion of orderly progress in scientific ideas. There was an actual resurrection by William Thomson (Lord Kelvin) of the old idea that atoms and molecules, with empty space between them, do not exist. Thomson wrote an unpublished manuscript in 1858 containing the "doctrine of the Universal Plenum," which viewed matter in terms of motions or eddies within the plenum, much as Descartes had done two centuries earlier. In 1867 Thomson reiterated and extended these ideas in a published paper "On Vortex Atoms."

The origin of this unexpected development is related to one of the great *progressive* developments of the same period, the creation of "thermodynamics." This new discipline of physics included a definition of energy and the formulation of the first and the second laws of thermodynamics, the universal rules that govern transformation of one kind of energy into another. A critical part of this development was the demise of the caloric theory of heat. Heat was no longer imagined to be an "element" (*caloric*), as it had been in Lavoisier's time, but was now known to be a manifestation of rapid motion in matter, a part of "energy" rather than a substance.

Thomson's vortex atoms arose from an "overshoot" of this idea. Perhaps not only heat, but matter as well, could be viewed as "motion"? The possibility appeared to be bolstered by Maxwell's theory of the electromagnetic field, which represented electromagnetic rays (including light rays) as oscillations transmitted through space with high frequency, and at the time seemed to require that "space" be a plenum, a medium usually called the *ether* (or *aether*), possessing real entities capable of carrying out oscillatory motion. Though experiments designed to demonstrate the actual existence of this space-filling medium invariably produced negative results, belief in the theoretical need for it persisted into the early twentieth century.

Scientists working on problems unrelated to propagation of electromagnetic radiation were, however, mostly able to ignore this problem, and pragmatically accepted the existence of particulate atoms and/or molecules, with "vacuum" between them. The common and uncomplicated reaction to the "heat is motion" idea was to assume that the discrete particles of matter were doing the moving. This led to the development of a highly successful new branch of theoretical physics called the *kinetic theory of gases*. It sought to provide a three-pronged link between *molecular* properties ("mass," "velocity"), *measurable* parameters ("pressure," "heat content"), and the abstract formalities of the laws of thermodynamics. Rigorous mathematical equations related these different conceptions of gaseous matter to each other, all based on the idea of myriads of molecules, confined to a limited space, but moving at high velocity within that space, helter-skelter in different directions, bouncing off each other and off the walls of their container. The success of the theory lay in the fact that it worked. No inconsistencies arose. (Other more approximate statistical theories, derived from the theory for gases, were able to link the properties of liquids and solids too, at least in principle, to their assumed constituent molecular particles.)

Chemists, too, tended to conduct their business within the framework of a strictly particulate view of matter. The kinds of molecular formulas drawn in figure 11 presuppose the existence of discrete atoms, and their combination with each other in integral ratios. Moreover, the number of known elements had grown, and therewith the number of distinctly different atoms. The earlier uncertainties in the relative combining weights of the atoms had all but disappeared, giving rise to an internationally accepted table of "relative atomic weights." Table 3 shows part of a very recent version, but, within the accuracy of a single decimal place, the 1890 values were not significantly different.

But belief in the reality of molecules did not mean that one yet knew much about their physical characteristics. In particular, the simplest of all questions—"How small is a molecule?"—was still unanswered. It was possible to have a clear mental picture of the

Table 3. Modern Relative Weights of the Atoms (only the first 26 elements are shown)

Element	Symbol	Relative atomic mass	Element	Symbol	Relative atomic mass
Hydrogen	H	1.0	Silicon	Si	28.1
(Helium	He	4.0)[1]	Phosphorus	P	31.0
Lithium	Li	6.9	Sulfur	S	32.1
Beryllium	Be	9.0	Chlorine	Cl	35.5
Boron	B	10.8	(Argon	Ar	39.9)[1]
Carbon	C	12.0	Potassium	K	39.1
Nitrogen	N	14.0	Calcium	Ca	40.1
Oxygen	O	16.0	Scandium	Sc	45.0
Fluorine	F	19.0	Titanium	Ti	47.9
(Neon	Ne	20.2)[1]	Vanadium	V	50.9
Sodium	Na	23.0	Chromium	Cr	52.0
Magnesium	Mg	24.3	Manganese	Mn	54.9
Aluminum	Al	27.0	Iron	Fe	55.9

[1] Elements in brackets are the inert gas elements, and they had not yet been discovered at the time of the surface monolayer experiments of Lord Rayleigh and Miss Pockels.

physical states of matter, composed of molecules with varying degrees of freedom of motion, without knowing the actual size of individual molecules. Chemists could confidently draw structural formulas for molecules, showing precisely how atoms were linked to each other without knowing how much space a single atom or the whole molecule might occupy. Likewise, chemists could confidently calculate how much A and B needed to be weighed out to make a compound (AB, AB_2, etc.) purely on the basis of the relative atomic weights of table 3, without knowing how much a single atom or molecule might weigh.

Even the kinetic theory of gases could not provide an accurate measure of molecular size, although its avowed purpose was to provide mathematically rigorous links between molecular properties and observable phenomena, such as the relation between the temperature, pressure, and volume of a gas. The simplest and most elegant part of the kinetic theory, which accounted perfectly for the *limiting* behavior of gases at low pressures, did not explicitly involve molecular size. The equations were derivable in their entirety by assuming only that colliding gas particles must be small in comparison with the distance traveled between collisions. A more exact definition of size was not necessary.

Compression of a gas of course leads to closer approach of the molecules to each other, and the assumptions on which the simple limiting equations were based must break down. But theoretical equations that could be applied above the low pressure limit were still crude, and, understandably, the size estimates made on this basis were quite uncertain. They did have one important effect, to demonstrate that molecular particles had real finite size and could not be dimensionless points, so that the Democritus/Boscovich debate was resolved—in favor of Democritus.

This situation made it logical for Lord Rayleigh to resurrect Franklin's experiment once he learned about it, as will be related in the following chapter. "In view of the great interest which attaches to the determination of molecular magnitudes," he wrote, "the matter seemed well worthy of investigation." The result he obtained—presumably one dimension of a single olive oil molecule—was quite close to the estimate we made in the last chapter on the basis of Franklin's visual observation: 16×10^{-8} cm as compared to 10×10^{-8} cm. The result was universally welcome, not only because it was much more precise than estimates based on kinetic molecular theory, but also because it was a *direct* measurement, dependent on no complex mathematical equations. Even Thomson was pleased and convinced. He wrote to Rayleigh on March 30, 1890: "I am delighted with your 16×10^{-8} film of oil on water. Think next of the calming effect of oil on water."*

4. Avogadro's Number

When the question is asked, "How small is a molecule?", it can be answered by giving the molecule's mass or weight instead of using a "length," as we have done so far. Mass or weight is actually preferable if one wants the scale of molecular size to be represented by a single number, because "molecular length" is an ambiguous term without some accompanying reference to molecular shape. Are molecules symmetrical? Is an olive oil molecule a perfect sphere, the measured dimension being its diameter? Or (more likely) are molecules asymmetric? If so, then the one dimension that is measured by the thickness of an oil film—which we might call "height"—could be larger or smaller than the "width."

Mass and dimensions are of course intimately connected: the

* *Note about molecular dimensions*. Rayleigh and his contemporaries expressed molecular dimensions in "micromillimeters," that is, units of 10^{-7} cm. At the beginning of the twentieth century "Ångstrom units," equal to 10^{8}cm, became the standard, and the symbol "Å" was used to avoid the necessity for an exponential notation. Quite recently it has been decided that "nanometers" (10^{-9} cm) should be the preferred unit. I shall use 10^{8}cm here and in later chapters, retaining the exponential notation to avoid any possibility of confusion.

dimensions determine molecular volume, and the *density* of the substance then determines mass,

$$\text{mass} = \text{density} \times \text{volume}$$

Using the dimension of 10×10^{-8} cm, for example, that would have been obtained from Franklin's oil film thickness if he had made the calculation, together with the density of olive oil (a little less than that of water), this equation tells us that an olive oil molecule has an actual mass of between 10^{-22} and 10^{-21} gram, the exact value depending on our assumptions about molecular shape. That's an even more dramatic number than the dimension itself in telling us how small real molecules are!*

The reason for bringing up this subject is that the conventional way that molecular size is today introduced to beginning students of chemistry is not through molecular dimensions, but through molecular mass, and, more particularly, through *Avogadro's Number*. This number, named after Amedeo Avogadro, the scholarly schoolmaster from Vercelli in Tuscany, relates real atomic masses to the *relative* atomic masses listed in table 3,

$$\text{Avogadro's Number} = \text{relative mass} / \text{true mass}$$

* It should be noted, for the sake of the nontechnical reader, that calculations of this kind are very easy to make and require no special training. Take the measured dimension of 10×10^{-8} cm, for example, and imagine the olive oil molecule to be spherical. A sphere with a diameter of 10×10^{-8} cm has a volume of 5.2×10^{-22} cubic centimeters (cm^3), and that would then be the molecular volume. Likewise, if we imagine olive oil to have rectangular boxlike molecules, with dimensions of, say, $(10 \times 10^{-8}$ cm$) \times (5 \times 10^{-8}$ cm$) \times (3 \times 10^{-8}$ cm$)$, then the molecular volume would be smaller than for the sphere, namely 1.5×10^{-22} cm^3. Assuming a perfect cube with each side 10×10^{-8} cm long, this would lead to a larger volume, 1.0×10^{-21} cm^3. The density of olive oil is about 0.9 gram per cm^3, so that the molecular masses that correspond to these volumes are 4.7×10^{-22}, 1.4×10^{-22} and 9×10^{-22} gram, respectively.

Avogadro himself did not measure this number nor did he even create the concept. It was named in his honor because he solved the riddle of the gaseous elements, understanding that many of these elements exist in nature, not as single atoms, but as diatomic molecules. Until this was understood, self-consistent values for atomic masses, as given in table 3, could not be assigned.

Since the mass of a molecule is the sum of the masses of its atoms, Avogadro's Number is all we need to obtain the *true mass of any molecule*, provided we know its chemical formula. The olive oil molecule, for example (Fig. 12), contains 57 carbon atoms, 104 hydrogen atoms and 6 oxygen atoms, for a total mass on the *relative* scale of 884. We must divide by Avogadro's Number (which is now known to have the value of 6×10^{23}) to obtain the actual molecular mass. The result (the *true* mass) is 1.5×10^{-21} gram, quite close to the estimates based on molecular dimensions.

The value of Avogadro's Number was not known in 1890—it was not determined with accuracy until 1913. Nor was there any known way to measure the mass or weight of a single molecule directly. The experimental oil film thickness gave the best number available for the scale of molecular size, and it is readily convertible to a molecular mass if that is what is wanted.

5. Note on the Birth of Modern Physics

Just about the same time that the world of physics was finally accepting the existence of atoms and molecules and coming closer to defining their masses and dimensions, momentous discoveries began to be made to upset the basic notion that atoms are indivisible and indestructible. It was the birth of so-called "modern" physics, the addition of an entirely new dimension to our perception of matter and of the cosmos within which it resides. We take note of it here, firstly because we cannot let an event of this importance simply pass by without asking how it affects the *old* conceptions of atoms and molecules, which in fact it hardly does at all, and,

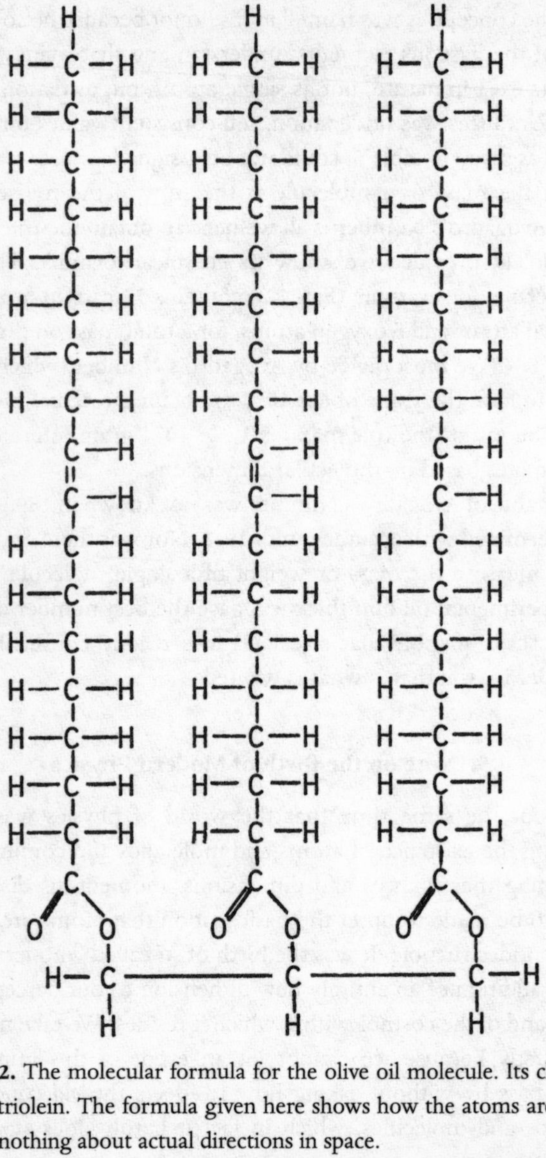

Figure 12. The molecular formula for the olive oil molecule. Its chemical name is triolein. The formula given here shows how the atoms are linked but says nothing about actual directions in space.

secondly, because the effect on established scientists of the time is interesting. Did they embrace it, oppose it, ignore it?

Here is an admittedly overcondensed summary. Modern physics began in 1895 with Roentgen's discovery of X rays. Then in 1896 came Becquerel's discovery of natural radioactivity, and in 1897 J. J. Thomson's discovery of the electron. Along with these essentially qualitative discoveries came startlingly new intellectual theoretical concepts: Planck's proposal for the quantization of energy in 1901, Einstein's first paper on relativity in 1905. Then, between 1911 and 1913, came the elucidation of atomic infrastructure by Rutherford and Bohr. From here on subatomic physics dominated almost all of physics, and thirty years later subnuclear physics took over from that, with the atomic bomb as its best-known public manifestation.

These advances are sometimes presented to the lay reader as a kind of revolution. Classical physics and chemistry are said to be dead, shown to be wrong, *replaced* by the "modern" physics and chemistry. Ptolemy had been wrong about the planets; Copernicus had been right. Aristotle had been wrong about gravity, and was superseded by Newton. Now Newton has been proved to be wrong, and Einstein is right.

Nothing could be further from the truth! Classical physics is alive and well, provided only that we don't apply it to travel at the speed of light. Classical chemistry, based on reactions in which atoms are preserved intact, is equally alive. The new chemistry impinges only when one considers the origin of radioactivity or processes within a nuclear reactor, where transmutations are deliberately fostered. Molecular biology, the chemistry of the living cell, to which we shall be moving our attention later in this book, is benignly classical, so much so that it is usually "explained" to students and the public by means of molecular models for DNA, RNA, and proteins that are assembled from scaled-up, hard, and assuredly indestructible plastic "atoms." (See Fig. 13.)

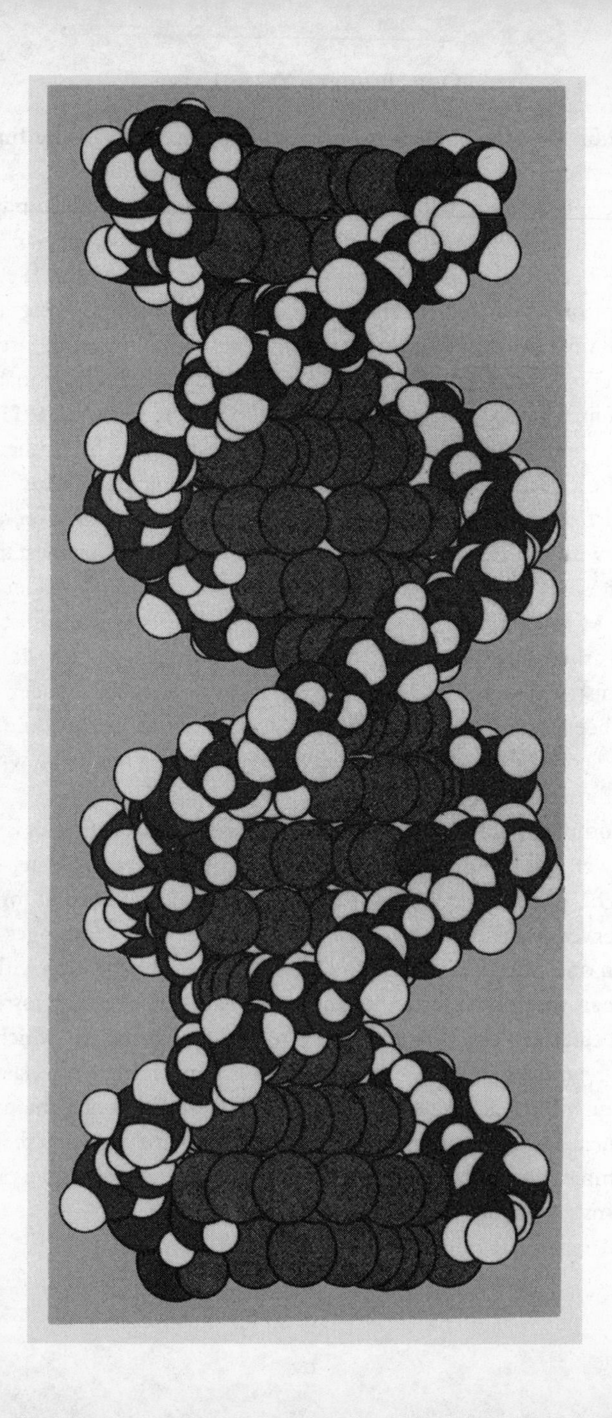

6. Stilling Waves at Sea

A letter from Lord Kelvin to Lord Rayleigh was quoted earlier, in which Kelvin congratulates Rayleigh on his oil film measurement and suggests that he might now "think" about the calming effect of oil on water. There was in fact renewed interest during the 1880s in the use of oil for stilling of the waves. Some of the accounts are more enthusiastic than Ben Franklin had been after his disappointing failure at Portsmouth Harbor (chapter 7).

One highly publicized effort came from John Shields, proprietor of a linen mill in Perth (Scotland). His intended method (announced in a preliminary patent) was to install undersea pipes in harbors, from which oil could be discharged when necessary. Tests were made in 1882 at Peterhead Harbor, on the northeast Scottish coast, and later at Aberdeen. The tests were successful, but the beneficial results were of short duration, and it was readily seen that a permanent installation and a supply of oil for continuous use through a long storm would very likely be unduly expensive.

Devices designed to be used by individual boats or small ships promised to be more practical, and the United States Hydrographic Office in Washington invited ship captains to send in reports of their experiences. Reports were published in the monthly *Pilot Chart* and collected together in separately published pamphlets under the title *The Use of Oil to Lessen the Dangerous Effect of Heavy Seas*.

Figure 13. A model for a part of the DNA double helix, built from indestructible plastic "atoms." The size of each kind of model atom is directly proportional to the true atomic diameter (usually on a scale of 1 cm per 10^{-8} cm of actual size). The faces where atoms combine with one another are sheared off to reflect the actual shrinkage that is known to occur when electrons are shared between atoms (see chapter 14). This shrinkage is a quantum chemical effect, the only place in this picture where "modern" physics intrudes upon an otherwise wholly classical conception. (See Fig. 11 for similarly constructed models of two smaller molecules.)

The most successful technique was to use a sturdy bag made of sailcloth and filled with oil-saturated oakum stuffing, "oakum" being a fibrous material made by hand from worn-out ropes. The bags were pierced at the bottom by many small holes. One or two such bags could be suspended from a boat's bow or stern, depending on exact weather conditions. Because the exit holes were small, the rate of efflux of the oil was also slow, only two or three liters per hour, but that was enough. (Remember, one teaspoon covered half an acre.) The slow rate of efflux made the method economical, permitting a ship to carry a supply of oil sufficient for several storms.

The reports of successful use of oil bags in Hydrographic Office pamphlets issued in 1886 and 1887 came from many places and from vessels of many nations: an Icelandic steamer in the North Atlantic, a United States schooner on its way to the Aleutian Islands, a lifeboat launched from a wrecked ship near Rio de Janeiro. All reports, from both harbor applications and single ship use, draw attention to the fact that oils are not all equal. Fish or seal oil was reported as the best, olive oil was less efficacious, and mineral oil, used alone, was (not surprisingly) practically useless.

Admiral Cloué, veteran of many campaigns in the French navy, and navy minister from 1880 to 1881 in the cabinet of Premier Jules Ferry, summarized the Washington reports in 1887, with an article in the *Comptes Rendus* of the French Academy. He urged France to use the method in the interest of safety. He also made a rough estimate (not as far out of line as one might expect) of how thin a film of oil appeared to be able to maintain the wave-stilling effect, basing his calculation on a ship's speed of travel, and on the minimal rate of oil efflux reported as being needed.

CHAPTER TEN

Lord Rayleigh

1. Amateurs in Science

THE word "amateur" has two different meanings. The first refers to someone who does what he does out of love, and not for profit. It implies no lack of professional skills or dedication. The second meaning, the more usual one today, is more explicitly the antithesis of "professional." It calls forth the image of a "dabbler," who does lack the skills and educational background of the professional.

The two "amateurs" we shall meet in this and the following chapter are opposites in this respect. Lord Rayleigh, a "gentleman scientist," wealthy by virtue of inheritance, has all the advantages of education and of association with equally educated peers. Agnes Pockels, on the other hand, without formal education, was by all appearances just a "dabbler" in her kitchen.

The "gentleman scientist" is not uncommon in the history of science, especially in Great Britain. One of the most celebrated lived in the sixteenth century—John Napier, Laird of Merchiston (in Scotland), the inventor of logarithms. He was a landowner, lord of a manor, with a good income from his crops and cattle. He was active in local and national affairs and a vigorous protagonist for the Protestants against the Papists. But his hobby was mathematics, and it gained him a prominent place in history. He not only "invented" logarithms, but undertook the laborious task of computing the world's first table of logarithms, something that required more than twenty years of his life. Adaptations of Napier's logarithms became the universal tool for numerical calculations from the day of first publication (in 1614) until the advent of commercial calculating machines in the middle of the twentieth century and, most recently, the advent of electronic computers.*

Other famous names in the "gentleman scientist" category include (in the seventeenth century) Robert Boyle, a son of the Earl of Cork in Ireland, inventor of a practical vacuum pump and creative investigator of the laws of expansion and compression of gases. In the nineteenth century there was Charles Darwin, who dropped out of medical school because the work was too hard, certain in the knowledge that the inheritance he would receive from his father (who *was* a doctor) would support him and his family in comfort for all his life. Late in the same century, there is a famous American example, J. Willard Gibbs, unsalaried professor at Yale University for many years (see brief biography in chapter 12). There are examples even in the twentieth century, such as King Gustavus VI of Sweden, who made significant contributions to archaeology and botany.

There have also been a few "lady scientists" burning with the same desire to know and understand the natural world around us. A famous example here is the Marquise du Châtelet, wife of a wealthy Lorraine landowner and also mistress to Voltaire, who made the definitive translation of Newton's *Principia* into French in 1656. (Her editor said it was even better than Newton's original Latin text because she clarified certain obscure passages.)

Benjamin Franklin and some of his friends in the eighteenth century were of course also true amateurs, devoting parts of their

*The generation now growing up, familiar with computers from their earliest days in school, will never be able to appreciate the indispensability of logarithms and slide rules—logarithmic scales etched on wood or metal. They were the only calculating aids available to Lord Rayleigh and Miss Pockels, or to any scientist for several decades afterward. Rayleigh, in fact, did not use even a slide rule, but relied exclusively on logarithms, filling page after page of his notebooks with patient manual calculations.

lives to science but earning their livelihood in other ways. They were not quite in the same category as the "gentleman scientists" because they were self-made men—wealthy by dint of their own previous labors or business acumen, not by inheritance. The most famous of Franklin's friends, Joseph Priestley, was not wealthy at all. He was an amateur in the sense that science was never a paying "job" for him, but he was always too poor to buy his own equipment and needed patrons to support his work.

But Agnes Pockels, as I said above, belonged to a different category. She was not abjectly poor—her family lived on her father's pension—but she was poor enough so that her life was by necessity one of drudgery in the kitchen, cooking the meals and washing up afterward. What were the inner resources that fed her desire to be a scientist? How could she, in spite of her lowly position and lack of formal education, become a *successful* scientist, acknowledged as such by her more worldly professional peers?

2. Biographical Sketch. Physics as a Pleasure

Lord Rayleigh (John William Strutt) was in a remarkably privileged position. It is not just that he was *born* to high rank and abundant wealth. He rose even higher in society by his lifelong friendship with Arthur Balfour and his marriage to Evelyn Balfour, Arthur's sister. The Balfours were niece and nephew of the Marquis of Salisbury, the most powerful man in British politics—at least on the Conservative side. Salisbury was Foreign Secretary in Disraeli's cabinet. He himself became leader of the party when Disraeli died in 1881, and Prime Minister whenever the party was in power. He served three terms in that capacity, for a total of fourteen years. When he retired in 1902 he selected his nephew Arthur as successor, and the prime ministership thus remained "in the family" for several more years.

Rayleigh stayed at the Balfour residence whenever he was in London. For many years the residence was at 10 Downing Street, because Arthur Balfour was First Lord of the Treasury under Salisbury, and 10 Downing Street was then the official residence of the First Lord, and not, as it is now, the residence of the Prime Minister.

It is thus quite clear that Rayleigh had more than the usual scope for personal satisfaction within the conventional framework of his station in life. He could have chosen the life of an English country squire (as his father had done): to maintain his estates, treat his tenants well, and perhaps engage in gentlemanly sports. Alternatively, he could have used his family connections to get some official administrative position or as an aid to a political career; a seat in the House of Lords was his as birthright even without the family connections. Instead, Rayleigh chose to devote himself wholly to science. He presumably did it because he liked to do science, got personal satisfaction from doing experiments and from mathematical analysis of physical problems. There is no hint of expectations of glory. When King Edward VII conferred the Order of Merit on Rayleigh in 1902, Rayleigh's son reports that his father responded to a toast by saying that "the only merit of which he personally was conscious was that of having pleased himself by his studies, and any results that may have been due to his researches were owing to the fact that it had been a pleasure to him to become a physicist." In fact, Rayleigh was altogether a singularly modest man, averse to ostentation, even in his personal appearance. He paid no attention to fashion and always wore his clothes till they were threadbare.

The Strutt family, though relative newcomers to the ranks of British nobility, had been landed gentry for many generations before that, owning well managed and prosperous estates at Terling, near Colchester, in the county of Essex. "Lord Rayleigh's Farms" is to this day a bustling and presumably profitable business. For a time they sold dairy products directly to the public through a chain of retail outlets in London. The retail stores were sold in 1928, but the farm's lorries, delivering produce wholesale to food shops, are still a common sight in London. And Terling Place remains an impressive country mansion, with admirable ornamental trees and flower gardens. In the modern fashion the gardens are open to the public one or more Sundays each year, and the servants provide afternoon tea for visitors at two pounds a head.

The family title was acquired by Lord Rayleigh's grandfather, a Tory M.P. and a colonel of the West Essex regiment. He was offered a hereditary peerage at the coronation of George IV in 1820, apparently for no reason other than long and faithful service to the crown—at least no conspicuous deeds of brilliance or valor are recorded. He accepted the honor for his wife and heirs, but (for unknown reason) not for himself. The name "Rayleigh" was chosen for the seat of the barony, after the name of a nearby village.

The scientific "Lord Rayleigh" was the 3rd baron. He was sickly as a boy and unathletic. His poor health interfered with his early education. For example, he had a mild case of smallpox and a more serious bout with whooping cough shortly after he enrolled at Eton, and he had to be withdrawn from the school after only one semester. Fortunately his parents saw to it that his schooling was continued in smaller private schools. Rayleigh's mathematial and scientific interests surfaced during these early school years and were soon recognized by parents and mentors.

Rayleigh enrolled at Trinity College, Cambridge, in 1861. The college had by then recovered from the doldrums described earlier. Rayleigh was not a conspicuously good student initially, but by 1865 he had emerged to top rank. That year he trained strenuously for the mathematical tripos examinations, which tested the ability to think mathematically about physical models (made as devilishly difficult as the examiners were able to make them), but required no factual knowledge about physical experiments. Rayleigh succeeded in winning the coveted senior wrangler position. Neither Maxwell nor Lord Kelvin, among his predecessors of comparable or even greater ultimate distinction, had been able to win that prize—both had been only "second wrangler" in their tripos years.

Rayleigh became a Trinity College Fellow in 1866, and it is amusing to note that his father was not enthusiastic, thinking that the appointment might be a little below his station in life and might suggest an intent to neglect his proper duties as the lord of a large estate. In any case, Rayleigh had to resign his position in 1871, when he married, because Fellowships were strictly limited to bachelors; this restriction was not abolished until 1882. (Before 1871 there had been the even more startling rule that Fellows must be members of the clergy—new Fellows had to go through the formality of taking Holy Orders!)

Sickness intervened again shortly after his marriage, when Rayleigh contracted rheumatic fever. Sir William Jenner (physician to Queen Victoria and recognized authority on fevers) was one of the physicians who attended him, and his prognostication was gloomy. Recuperation, however, slowly came. The cure included stimulants that would have been outside the reach of anyone less well connected: a three-month-long trip up the Nile in a chartered houseboat, and, on his return to Cairo, a land-tour of Greece, where the Rayleighs were joined by Arthur Balfour. It is characteristic of Rayleigh that he was not idle during these travels. On the contrary, he worked all morning in his ship cabin throughout the Nile voyage and completed one of his major works, *The Theory of Sound*, there.

It is also characteristic of Rayleigh that *The Theory of Sound* (two volumes and 984 printed pages in a modern paperback format) was entirely handwritten on the backs of old student examination papers! Rayleigh was a very economical person. (All of Rayleigh's publications and correspondence were handwritten. He never employed a secretary.)

Rayleigh succeeded to the family title ("Baron Rayleigh") when his father died in 1873, but he soon turned the management of the estate and farm business over to his younger brother Edward. He converted one wing of the Terling mansion into a laboratory and purchased all the equipment that he needed for it, including an expensive telescope. During the rest of his life, more than forty years, he spent most of his time there. He lived in Cambridge twenty weeks out of the year during the short period when he was Cavendish Professor, and he went to London perhaps once a week, to meetings of the Royal Society, etc., but most of his days were spent at Terling, and they followed, year after year, a regular domestic routine centered on his work. Breakfast was at 9 a.m., and letters were read at the table and answered in his study after breakfast. Then journals were read, calculations made, and papers written. After lunch he might spend some time with his children, but laboratory work occupied most of the afternoon and sometimes continued after dinner. Meals had to be strictly on time, presumably no problem with his wife as manager and a large household staff. Rayleigh produced nearly five hundred papers in his lifetime, the last one completed just five days before his death at age seventy-six.

Rayleigh appears to have had little need for recreation. There were often guests at Terling (especially members of the Balfour family), but their perturbation on his routine was probably small. One interest outside physics was psychic phenomena. Rayleigh actively participated in a number of spiritualist seances, and a few took place at Terling itself. Rayleigh was a member of the Society for Psychic Research, and, near the end of his life, served as the society's president for one year.

In 1879 James Clerk Maxwell died at the early age of forty-eight. He had been the first Cavendish Professor at Cambridge, as was related in the preceding chapter. William Thomson (later Lord Kelvin), the original first choice for the post, was first choice as replacement, but he again declined, and the position was then offered to Rayleigh. An agricultural depression had hit England that year, and Rayleigh's income from his estate fell significantly; with this inducement he accepted, though without enthusiasm. He wrote to his mother that neither he nor his wife relished the idea of living in Cambridge, but they thought they ought to take the chair for three or four years "if they can get no one else fit for the post."

In another letter to her a couple of weeks later he said "it is open to question whether in any case I could have gone on living as hitherto, as my financial difficulties seem rather to increase." It was the only normal "full-time" paying job that Rayleigh ever held, and he resigned it thankfully in 1884, his farm income by then having returned to normal.

Much later, during the last ten years of his life, Rayleigh was appointed Chancellor of Cambridge University, a position of great eminence, no salary, and few duties. It interfered little with his scientific work at Terling.

In 1887 Rayleigh accepted another professorship, but this time without reluctance. The post was that of professor of natural philosophy at the Royal Institution, a marvelous establishment in London, founded as a philanthropic laboratory in 1799 by Benjamin Thompson (Count Rumford) and others. It was intended to serve both as an educational institution and as a laboratory to develop technology that would advance the condition of the poorer classes of society. A principal part of the educational function of the Royal Institution was (and remains) to educate the general public, and Rayleigh's duties as professor were to give six public lectures a year, on six Saturday afternoons just before Easter. He was also required to give one of the fashionable (and more advanced) Friday night lecture-demonstrations at the Institution, which were attended by all the scientific elite in London, and were in part designed to attract wealthy and influential "patrons." As we shall see later, Rayleigh received much inspiration for his research from these lectures. This is not unusual: the need to prepare a relatively broad *elementary* lecture is often an inducement to make intellectual connections that would not occur in straight "linear" thinking about a previously defined research goal.

Apart from his strictly scientific occupations, research, and lecturing, Rayleigh had a strong sense of civic duty and held a number of public service positions. One was a position normally considered to be rather time-consuming, the position of secretary of the Royal

Society, which he held from 1885 to 1896, and took very seriously. His publication record during those years (which included the discovery of argon and the oil-on-water experiments) indicates, however, that he did not allow the position to interfere too drastically with his work at Terling. (In modern times the position of secretary of the Royal Society is considered so valuable a public service that it has become traditional to award a knighthood to its tenants.)

Naturally, Rayleigh was a member of the House of Lords by virtue of his rank, and he was also on intimate terms with figures in the upper echelons of government. He took considerable interest in political affairs, generally holding archconservative opinions, but he rarely exercised his privilege of expressing them in the public forum. In fact, he participated in parliamentary debate only when

Figure 14. Lord Rayleigh. Royal Society portrait, painted by Sir George Reid in 1903.

he felt himself especially qualified to do so. One recorded occasion was in 1876, when he proposed an amendment to the Cruelty to Animals Bill in the interests of physiological research, and another was in 1886, when he participated in debate concerning the Electric Lighting Act. He held the purely political post of Lord Lieutenant of the County of Essex for many years, and in Westminster he was a member of the explosives committee of the War Office. Around 1900 he became an active member (de facto chairman) of the important committee that created and later supervised the National Physical Laboratory in Teddington, near London.

Rayleigh died of a heart attack on June 30, 1919, having remained fully active to the very end. He had dined at Buckingham Palace on December 27, 1918, at the royal banquet in honor of President Woodrow Wilson. He had delivered the presidential address to the Society for Psychical Research on April 11, 1919. He finished dictating a scientific paper for his wife to write on June 25, 1919.

He was buried at Terling. The funeral was attended by the principal officers of Cambridge University and of the Royal Society. Two years later a memorial tablet with a medallion portrait was put up in Westminster Abbey.

For the sake of completeness, it should be mentioned that Lord Rayleigh has been criticized for remaining aloof from the revolutionary developments of modern physics that were taking place all around him, and, more generally, for failing to be "a boldly imaginative scientist who would initiate a wholly new idea." The factual part of this statement cannot be denied. Rayleigh's scientific outlook was, like his outlook on life in general, rather conservative. Whether he is to be criticized for this is, however, another matter. Sir J. J. Thomson, in the address he gave at the dedication of the memorial tablet in Westminster Abbey had this to say:

> There are some great men of science whose charm consists in having said the first word on a subject, in having introduced

some new idea which has proved fruitful; there are others whose charm consists perhaps in having said the last word on the subject, and who have reduced the subject to logical consistency and clearness. I think by temperament Lord Rayleigh really belonged to the second group.

3. Why the Sky Is Blue, and Other Profound Matters

Rayleigh's lasting fame among physicists rests on his application of complex mathematics to a host of problems in theoretical physics. The monumental *Theory of Sound*, written early in his career, is one example. Much of his later work concerned the radiation of light waves, quite different from sound because the radiated oscillations are transverse in light (perpendicular to the direction of propagation), whereas they are longitudinal for sound (in the same direction as the direction of propagation). Included among his papers on optics and light are theoretical analyses of interference and of light scattering that have remained standard texts up to the present time.

Another still frequently used theoretical analysis is the problem of the "random walk" or "random flight." Suppose that *n* steps are taken, of specified lengths (not necessarily all the same) but completely random in direction. What will be the distance from the starting point after the *n* steps? The result would of course be different every time a trial is made: sometimes one would have doubled back almost to the starting point, other times one could be quite far away. However, the word "random" in the statement of the problem specifies a lack of preferred direction, and this makes the problem amenable to a pure statistical analysis. Rayleigh carried out this analysis, and his result gives the probability that the overall beginning-to-end distance will fall within any given range. The result finds application to all kinds of practical problems. In polymer chemistry, for example, Rayleigh's equation (in three dimensions) forms the basis for calculating the distance between the beginning and end of a chain polymer molecule.

In spite of this reliance on mathematics as a tool, not only for solving problems but often even for defining where a problem exists, Rayleigh had a knack for appreciating and addressing questions of popular appeal. One outcome of his work on light scattering, for example, was an explanation for an age-old mystery. Why is the sky blue? As Rayleigh points out in his *Encyclopaedia Britannica* article "Sky," discrepant and unsupportable views on this question had been held even by writers of good reputation. The true reason is the result of sky light being *scattered* light. Without scattering, we would receive light from the sun only on a direct path, and the sky around the sun would be black. But sunlight not on a direct path to us is scattered in all directions by the air molecules in our atmosphere, and Rayleigh's theoretical equations show that the scattering depends on the wavelength of the radiation, such that blue light should be scattered at higher intensity than the other components of the sun's spectrum. As Rayleigh points out, the theory explains "not only the blue of the zenith," but also "the brilliant orange and red coloration of the setting sun."

In the eyes of the general public Rayleigh's greatest scientific achievement is usually considered to be the discovery and isolation of argon. The discovery itself was accidental, in the sense that there was initially not the slightest intent to search for a new element. It arose instead out of Rayleigh's characteristic concern with a profound question of classical theoretical physics, and out of the application of painstaking, patient experimental work in the effort to solve the problem.

The theoretical question in this case is suggested by the relative weights of the atoms, as shown in table 3. Why are so many of these numbers so close to integral multiples of the weight of a hydrogen atom? Atoms were still the indestructible ultimate particles of matter at the time—no hint of subatomic structure as yet—so what could possibly be the origin of integral ratios? Rayleigh did not actually think about the question directly, but decided that the question called for some especially accurate experimental measurements, to determine actual atomic weight ratios with the greatest possible precision. The measurement of gas densities, exactly proportional by Avogadro's law to the masses of the constituent molecules, was the method he chose. Rayleigh had previously measured accurate densities for oxygen and hydrogen (the measured ratio was 15.882), but for various reasons he particularly wanted to know the corresponding ratio for oxygen and nitrogen, and therefore he set out to measure the gas density for the latter. The result proved to be the most dramatic event in Rayleigh's life, leading to the discovery of argon and to the award of a Nobel Prize.*

Rayleigh first determined the density of nitrogen derived from air, after all the more reactive components of air had been removed by standard methods. He then decided to check the result by using nitrogen from another source, namely from the decomposition of ammonia. The values obtained were not the same! The result was completely baffling, and Rayleigh reported it in a famous Letter to the weekly magazine *Nature* in 1892. In his own words. "The relative difference, amounting to about 1/1000 part, is small in itself; but it lies entirely outside the errors of experiment."

The cause of the discrepancy was found in a collaborative investigation between Lord Rayleigh and the chemist William Ramsay. The nitrogen sample obtained from atmospheric air, after everything known to be in air that was not nitrogen had been removed, still contained a contaminant, something not then known, even more inert than nitrogen, which Rayleigh and Ramsay called

* Rayleigh's decision to make these measurements can perhaps be considered an example of his disinclination to be "boldly imaginative." A more aggressive response to the existence of the problem might have been a purely theoretical one, to try to imagine a physical model that would *predict* integral ratios. In this case, of course, the conservative approach paid high dividends. The irony of it is that a sensible theoretical approach to the problem of the internal structure of atoms—including near-integral mass ratios—would hardly have been possible without knowledge of the existence of argon and other inert gas elements.

argon, after the Greek word for "idle." There were initially many skeptics, who could not believe that there could still be a previously unknown component of air after a century of intense investigation by top-notch scientists. Several implausible explanations were proposed that did not require assertion that a new element had been discovered, but Ramsay and Rayleigh themselves refuted all objections, eliminating any possible accidental origin for the denser component.

A couple of years later, Ramsay isolated a second inert gas, helium, and invited Rayleigh to collaborate with him again in its definitive characterization, but Rayleigh declined. Argon, helium, and the other inert gas elements proved to be of the utmost theoretical importance when the first models for the internal structures of atoms were proposed a decade or two later.

Ramsay and Rayleigh were awarded *separate* Nobel Prizes for the discovery of argon in 1904, Ramsay in chemistry and Rayleigh in physics. Rayleigh donated his entire cash award ($38,500, in 1904 dollars) to Cambridge University, to improve the Cavendish Laboratory and the university library.

4. Oil on Water on a Laboratory Scale

In 1890 Rayleigh repeated Ben Franklin's experiment of spreading oil on a water surface, measuring the area to which a given volume of oil would expand. Rayleigh reduced the experiment to laboratory scale (smaller amount of oil, smaller area), but otherwise the procedure was essentially the same. Rayleigh, of course, did calculate the thickness of the surface film, which Franklin had not done.

The modern commentator might wonder what would prompt Rayleigh, with his normal predilection for the mathematical side of physics, to do something so utterly simple as measuring the spreading of oil on water, an experiment where the only equation to be used would be the elementary equation relating the volume, area, and thickness of a portion of fluid. Part of the answer lies in the general state of physics, summarized in the previous chapter. There was (by 1890) virtually complete acceptance of the existence of molecules, but at the same time an acute awareness that nobody knew exactly how small they were. However, the more direct stimulus for Rayleigh to become involved in this question and to resort to this particular experiment in doing so can reasonably be conjectured to have come from Rayleigh's appointment as Professor of the Royal Institution and the intimate familiarity with the work of Thomas Young which he acquired as a result.

Thomas Young (1773–1829). Thomas Young was one of the most versatile and brilliant of all English men of science, though he was not generally accepted as such at the time. He knew a dozen languages; he was a practicing physician and published several medical treatises; he discovered optical interference and thereby effectively proved the wave theory of light at the expense of Newton's corpuscular theory; he was the first to explain the phenomenon of surface tension at a liquid surface; he generated theories of elasticity and of ocean tides; he was the genius who worked out the key to the Rosetta stone, one of the first Egyptian hieroglyphics to be deciphered. This last event stimulated him to become an expert Egyptologist, and he was invited to write what proved to be a landmark article on Egypt for the 1818 supplement to the fifth edition of the *Encyclopaedia Britannica*. (He had articles on "capillarity" and "chromatics" in the same series of supplements.)

As might be expected from this long and diverse list of accomplishments, Young's work often consisted (by his own admission) of "acute suggestions," rather than fully documented airtight proofs. Moreover, some of his papers were published anonymously. All this accounts for his somewhat ambiguous reputation, but it should be noted that his stature has grown with time; in the intervening almost two centuries most of his "acute suggestions" have proved to be correct.

Young had been the first Professor at the Royal Institution, from 1801 to 1803, and the lectures he gave during his tenure of the position were published in 1807 under the title, *Course of Lectures on Natural Philosophy*. Lord Rayleigh studied Young's book in preparation for his own lectures, and in 1892 he based his own lecture series directly on the book and even used for his demonstrations some of the apparatus that Young had designed and had left behind at the Institution. Young's influence on Rayleigh, however, extended far beyond the public lectures. Rayleigh refers to Young repeatedly when writing about his own research, often in the context of correcting others who had claimed that no one had ever done such and such, pointing out that Young had in fact done the very thing. Rayleigh's intellectual debt to Young was even officially recognized after his death, in the placement of the Rayleigh memorial tablet in Westminster Abbey. The tablet was placed in the north transept, symmetrically with the memorial for Young. In the words of the biography written by Rayleigh's son and successor to his title, this location "was thought to be specially appropriate in view of the intellectual affinity between the two men."

Young's work on surface tension is of special interest here, and was to Rayleigh, prompting the latter to write a long two-part general theoretical paper on surface forces, in 1890, the same year in which he made the oil-on-water measurements. It should be explained to the nontechnical reader that surface tension is a measure of the tenacity with which a liquid surface resists expansion, and thereby becomes a measure of the attractive forces between molecules in the surface. Surface tension is easily measured experimentally by several methods. One that the general reader will appreciate depends on the measurement of the size of the drops that emerge when a liquid passes through a vertical tube—the higher the surface tension, the larger the drops grow before they become detached from the tube. (The credit for offering the first rational explanation of surface tension is often given to Young's more famous French contemporary, Pierre Laplace, but Rayleigh went to some pains to correct that error and to assign at least equal credit to Young.)

An unexpected item in Young's theoretical analysis of surface forces (especially when the early date is considered) is a remarkably ingenious calculation of the probable size of an individual water molecule, based on the measured surface tension of water. This would surely have excited Rayleigh and stimulated him to think about other ways to use simple surface measurements to estimate molecular size. Rayleigh does not actually say that this was the intellectual pathway to his own measurement, but it is implied in the prefatory acknowledgments made by Rayleigh when he supervised the publication of the third volume of his collected work in 1901.

Rayleigh presented one version of his experiment at one of the public lecture-demonstrations at the Royal Institution. He knew about Franklin's work by this time and used olive oil for the experiment. He introduced the subject to his audience with an account of Franklin's enthusiasm for demonstrating the stilling of the waves. "A pond is necessary to illustrate the phenomena properly," Rayleigh said, "but we shall get an idea of it by means of this trough six feet long, containing water." Rayleigh had used organ bellows in his own laboratory to create an artificial wind, but no bellows were available in the lecture hall, and he used an electrically driven fan instead. The wind set up ripples, and they were smoothed by a drop of oil just as they had been on the pond.

Rayleigh did not have time to discuss the estimation of molecular size in his public lecture, but he published a parallel paper in the *Proceedings* of the Royal Society, in which he explicitly addressed that subject. In this case, for quantitation, he used the motion of camphor chips as a measuring device instead of waves. The chips move about spontaneously on an uncontaminated water surface, but motion ceases where the surface is covered by oil. Rayleigh added increasing amounts of oil to the surface of water in a circular dish with a diameter of about one meter (i.e., he used a fixed area) and then calculated the film thickness from the volume of oil that was "about enough" to "very nearly stop" all movement of the

Table 4. Thickness of Spread Olive Oil Film on Water

	Amount of oil added	Area at full expansion	Film thickness
Franklin (1774)	1 teaspoon (2 cm^3)	1/2 acre	10×10^{-8} cm
Rayleigh (1890)	0.8 mg (0.0009 cm^3)	5500 cm^2	16×10^{-8} cm
Pockels (1892)	1.0 mg (0.0011 cm^3)	8460 cm^2	13×10^{-8} cm
Langmuir (1917)[a]	–	–	13×10^{-8} cm

[a] Langmuir did not need to report total area and total applied mass because Avogadro's number was available to him, so that he could calculate molecular data directly. He was able to measure cross-sectional area as well as height above the surface. See table 5 in chapter 15.

camphor. The average result of several determinations (see table 4 for one example) was 16×10^{-8} cm, very close to the result calculated in the preceding chapter from Franklin's experiment.*

Table 4 also gives results obtained by Agnes Pockels and Irving Langmuir, whose lives and work will be discussed in subsequent chapters. It will suffice here to note that Agnes Pockels invented a far better experimental technique than Rayleigh had used, a technique which Rayleigh himself employed a few years later. At that time he gave a lower estimate (10×10^{-8} cm) for the olive oil molecular "height." It should be emphasized that the small discrepancies between these estimates at different times are unimportant here. A molecular dimension between 10 and 16×10^{-8} cm is correct by present-day criteria, and the degree of uncertainty, less than a

*G. van der Mensbrugge at the University of Ghent in Belgium, who was well known for studies of water surfaces in general, had published a sound theoretical basis for the motion of camphor chips on pure water surfaces. Henri Devaux, in France, began using the same method to study oil on water in 1888, but he attached his camphor to little tin "ships" with masts and flags, as illustrated in figure 1 in the Introduction. Devaux never used his experiments to estimate the thickness of the minimal covering film. In a review written in 1931 he gives complete credit for that idea to Rayleigh.

factor of 2, is far below the uncertainty that had existed before 1890. Many modern references cite Rayleigh's result as the first *reliable* measurement of an actual molecular dimension.

5. Noblesse Oblige

It has been mentioned that Rayleigh was not motivated in his work by dreams of personal glory. He embodied, in fact, all the elements of the Confucian ideal of virtue—in particular modesty, kindness, and generosity of soul. He lacked the competitive assertiveness of his friend and fellow-physicist Lord Kelvin, for example, or that of Irving Langmuir later on (see chapter 15). This does not mean that he was oblivious of the need to record priority, to give credit where credit is due, but it is a historical fact that he was selflessly concerned with other people's credit much more than he was with his own.

His efforts on behalf of the memory of Thomas Young have already been cited. The "Waterston affair" provides another example. When Rayleigh was secretary of the Royal Society, one of his duties was to make the initial decision on papers submitted for publication to the Royal Society journals. He had to decide whether a referee should be consulted, and to choose the referee if one was deemed necessary. The job also gave him access to old files and to make retrospective judgement of fair treatment, and this led to the unprecedented disinterment of a paper by a Scottish scientist, John James Waterston, who had already been dead for several years. Rayleigh discovered that Waterston had been the victim of inexcusably shabby treatment by the publications committee of the Royal Society nearly 50 years earlier. He had submitted a theoretical paper to the Society in 1845, for publication in the *Philosophical Transactions*, and this paper was an outstanding example of "boldly imaginative" science. It stated explicitly for the first time some of the most fundamental principles of the kinetic theory of gases—the very rocks upon which the theory is founded, namely that heat is a measure of the kinetic energy of motion of the gas molecules and that the pressure of a gas is due to the impact of molecules against the side of the container. The paper, however, was rejected by a referee. "The paper is nothing but nonsense, unfit even for reading before the Society," the referee wrote.

This was outrageously unjust, and not only in retrospect, for the view of heat as molecular motion was beginning to be speculatively entertained by other physicists at the same time. Waterston of course wanted to try to publish the rejected paper in another journal, but he ran afoul of a now-unbelievable rule, according to which the Royal Society had the right of ownership over any manuscript sent to it for publication. Acting on the basis of this rule, they had refused to return the manuscript to its author. Waterston had not made a copy and apparently was unwilling to write a new manuscript. When Rayleigh came upon a reference to this paper in a later Waterston publication, he began to look for and soon found the rejected manuscript in the Royal Society archives. He then sponsored its publication *in full* in the 1892 *Philosophical Transactions*, with an introduction by himself. (In case anyone wonders why Waterston had no copy, and was not prepared to rewrite, it might be relevant to note that the eventually published paper comprises eighty-eight pages of printed text. The manuscript, of course, had been handwritten.)

The most charming example of Rayleigh's benevolence, and one central to the story of oil on water, occurred in 1891, when one morning he received a letter from a lady in Braunschweig (Germany), unknown to him, without credentials or references. She was Agnes Pockels, the subject of the following chapter in this book. She had apparently been investigating oil films on water in her kitchen since 1881, but had not known whether there was sufficient interest in her work to warrant publication. When she read Rayleigh's paper on the subject, she realized not only that interest did indeed exist, but also that her own experimental technique was probably superior to Rayleigh's and even that her theoretical insight was at least on a par with Rayleigh's. She then wrote to Rayleigh, enclosing a paper that described her work. This led to an extraordinary publication in the journal *Nature*, reminiscent in character of the letters in the *Philosophical Transactions* a century earlier.

I reproduce verbatim the first three paragraphs. The first paragraph is from Lord Rayleigh to the editors of *Nature*, and it is dated March 2, 1891:

> I shall be obliged if you can find space for the accompanying translation of an interesting letter which I have received from a German lady, who with very homely appliances has arrived at valuable results respecting the behaviour of contaminated water surfaces. The earlier part of Miss Pockels' letter covers nearly the same ground as some of my own recent work, and in the main harmonizes with it. The later sections seem to me very suggestive, raising, if they do not fully answer, many important questions. I hope soon to find opportunity for repeating some of Miss Pockels' experiments.

This was simply signed "RAYLEIGH." The remainder of the publication is a full translation of Agnes Pockels's letter to Rayleigh.

> MY LORD,—Will you kindly excuse my venturing to trouble you with a German letter on a scientific subject? Having heard of the fruitful researches carried out by you last year on the hitherto little understood properties of water surfaces, I thought it might interest you to know of my own observations on the subject. For various reasons I am not in a position to publish them in scientific periodicals, and I therefore adopt this means of communicating to you the most important of them.
>
> First, I will describe a simple method, which I have employed for several years, for increasing or diminishing the surface of a liquid in any proportion, by which its purity may be altered at pleasure.

And the description of her method and results followed.

Who would have thought that a man as busy as Lord Rayleigh would do more than thank this lady for her interesting letter? Even if he later modified his own experimental method on the basis of her letter, as turned out to be the case, scientific etiquette required no more than an explicit footnote acknowledging his debt to her. Instead, Rayleigh mobilized all of his energies to secure publication of Agnes Pockels's work as quickly as possible.

Agnes Pockels's letter was dated January 10 and presumably was delivered two or three days later. It was written in German and needed to be translated before Rayleigh could read it with full comprehension—Lady Rayleigh is said to have helped with the translation. Rayleigh cannot have had previous knowledge of who Miss Pockels was, and would have needed to study the letter with great care to convince himself that the contents should be taken seriously, that this was not the writing of some crackpot. (In fact, he exchanged several letters with her to clarify questions that he had.) Yet Rayleigh's letter of transmittal to *Nature* is dated March 2, only six weeks after January 10. The publication date, incidentally, is March 12.

We marvel, as we did once before, at the speed of communication. Why does it take so long nowadays to process publications?

CHAPTER ELEVEN

Meticulous Miss Pockels

Fräulein Agnes Pockels . . . has taught us, not only to create cleanliness, but even to measure it.

Wilhelm Ostwald in an appreciative article in *Kolloid Zeitschrift*, on the occasion of Miss Pockels's seventieth birthday in 1932

THEY all call her "Miss Pockels," never just use her last name. Rayleigh does this and so does Langmuir, even in reviews written many years later. It was done out of natural gallantry, I suppose. Even among servants, only the maids were "Smith" or "Jones," but the cook was entitled to "Mrs. Bridges." Men, on the other hand, addressed each other by their surnames alone, even if they were close friends. Given names were never used among adults.

"Sexism!" some will complain, and they might even extend that accusation to me. I bring a woman into my book, they will say, and what does she do? *She cleans!* Where does she do her work? *In her kitchen!* But I only report the facts. It was Wilhelm Ostwald who wrote the quotation at the head of this section. It might have been worded more elegantly, but what he said is true. Agnes Pockels's important contribution to the development of surface chemistry was her simple method for keeping the surface free of contamination—quickly and repeatedly, before every measurement. And her experience in the kitchen, cleaning greasy pots, was undoubtedly her source of inspiration. In retrospect, the Pockels method seems obvious, but it had not occurred to anyone before, and nobody

Figure 15. Agnes Pockels. From an appreciative article in *Kolloid Zeitschrift* for 1932.

knows how long it might have taken for someone else to think of it.

It is of course strikingly clear, from Agnes Pockels's own statement, that the society in which she lived was grossly discriminatory against women, that it deprived her of the formal education she wanted, and kept her tied to stove and sink. (To this day Germany has far fewer woman scientists than many other European countries: England, France, and Poland, for example.) In spite of this, Agnes Pockels became a scientist of repute, an independent scientist, doing original work, in no way patronized by her more famous younger brother. Her work has stood the test of time and is quoted to this day. To me she is a hero! (Or should I say "heroine"?)

Wilhelm Ostwald persuaded Agnes Pockels to write an autobiographical summary for her seventieth birthday article, and it is a poignant document:

> I was born on February 14 1862 in Venice, where my father was at the time a captain in the Austrian military garrison. Both parents, however, were North Germans.

That is the simple way she begins her story. Venice was an Austrian province at the time, and both Agnes and her younger brother Friedrich were born in the province, Agnes in Venice itself and Friedrich in nearby Vicenza. The province was not a healthy place—ridden by mosquitos and malaria—and Agnes's father was among the victims. He received a medical discharge from the service in 1871, at the early age of forty-one, and the family then moved to Braunschweig, in north central Germany. Agnes attended the city "Mädchenschule" there for several years, whereas her brother of course attended the "Gymnasium," and, later, the technical high school. In her own words,

> I acquired already then a passionate interest for science, especially physics, and would gladly have studied it, but women had not yet the right to study, and later, as the first beginnings were made, the wishes of my parents prevented me from doing so. So I cannot lay claim to a doctor's degree.

She proceeded to educate herself by self-study, first by means of a single little textbook, later through books provided by her brother Friedrich when he became (in 1883) a doctoral student at Göttingen.

> However, I did not get very far in this way in the mathematical treatment of physics, so that, to my regret, I understand only little of theoretical things.

Friedrich subsequently became professor of physics at the University of Heidelberg, but that was many years later, too late to affect the events and work related in this chapter.

Agnes Pockels made her first observations on surface phenomena in 1881 (age nineteen) and in 1882 developed her simple device for changing the area of a surface layer by means of a wire or metal strip, "all with quite primitive means." She comments:

> Publication seemed at first out of the question. I did not even know whether my observations were already long known. The Göttingen physicists, to whom I communicated my results, showed no particular interest in them.

But she continued her experiments and in 1890 learned that the great Lord Rayleigh was doing similar research.

> My brother advised me to write to him, and Lord Rayleigh then helped me himself to achieve publication in *Nature*.

"Hereby encouraged," Miss Pockels went on to publish a total of fourteen articles over a period of thirty-five years. Her primitive measuring device, as Irving Langmuir wrote in 1939, "laid the foundation for nearly all modern work with films on water," and it remains a standard tool of surface chemists and physicists to this day. A picture of a modern trough is shown in figure 16.

The design of the trough is based on clear and purposeful thinking, comprehension translated directly into practical application. The primary purpose was to measure surface tension, the force between molecules at the water surface that resists extension of the surface, and to observe how contaminants affect it. She was of course restricted in her method of measurement: it had to be one in which the planar surface of the liquid in the trough was retained. Her choice was to use a tiny disk (initially a simple button) suspended horizontally in the surface and attached to a homemade balance. The weight required on the other balance arm, an amount that would just barely raise the small disk from the surface, was the quantitative measure of the tension.

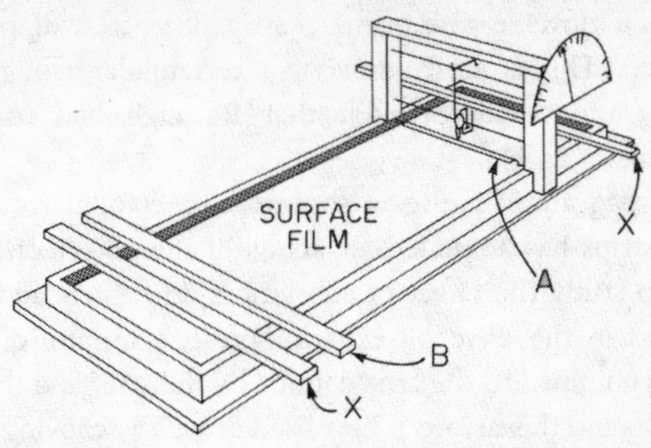

Figure 16. Diagram of a *Langmuir trough*. The basic design is essentially the same as that of the Pockels trough. *A* is a fixed barrier, *B* is a movable barrier by means of which the area of the surface film can be changed. The barriers labeled *X* are used to sweep the surface to keep it immaculately clean. In the Langmuir version (shown here) a pressure-measuring device is attached to *A* to measure the pressure change as the film is compressed to a smaller area. Agnes Pockels had measured surface tension instead by means of a tiny piece of metal attached to a standard chemical balance. (The two quantities—surface pressure and surface tension—are actually mathematically related, so that one could calculate one from the other. Surface pressure, measuring how hard one is pushing, is intuitively more appealing.)

It was common knowledge that surface tension measurements for supposedly pure water (by any technique) were difficult to reproduce. Dipping one's finger into the water reduced the surface tension, presumably because oil from the skin, like olive oil, spreads over the surface and does not dissolve in the bulk liquid. Other chance contaminants, such as dirt adsorbed from the air, though minute in amount, are also likely to be concentrated in the surface and to affect measurements to an extent quite out of proportion to the total amount present. But if contaminants that interfere with accurate measurement are concentrated in the surface layer, then they should be easy to remove by simply sweeping them away with a movable wire or metal strip, just barely dipping below the surface. This is easily done if a rectangular trough is used instead of the circular trough that Rayleigh had used in his experiments.

As the diagram in figure 16 shows, the rectangular design with movable strips has an added advantage if the object of an experiment is to study the effect of substances *deliberately* added to the surface—as in the olive oil experiment. The minute quantity of additive is confined to the area defined by the strip and the sides of the trough, and this area can be varied at will by moving strip B in the diagram. Agnes Pockels was able to make continuous (and *reversible*) changes in area in each individual experiment, whereas Rayleigh's apparatus had a *fixed* area and he had to begin all over again with a different amount of added substance for each new observation.

All of this was quite sophisticated for the times, and truly astonishing when Agnes Pockels's lack of formal education is taken into consideration. Rayleigh, when he returned to surface area measurements some years later, adopted the Pockels technique in place of his own procedure. But it was not only a matter of experimental technique. Agnes Pockels clearly understood that the primary effect of any surface contaminant was a direct effect on surface tension, and that the measurement of surface tension itself is therefore, even on theoretical grounds, the best method for studying the state of a surface. And this is the method that has been employed universally ever since.

Agnes Pockels's brother died in 1913, when he was less than fifty years old. His death and the war and Germany's general postwar turmoil effectively severed her regular contacts with the world of physics, though she did publish one more paper. At home, she began to blend in with her environment, meeting other ladies of her age for afternoon coffee and puzzle-solving. She became kind old "Tante Agnes" and kept to herself any memories she had of once being more than that.

But the world did not forget, and Agnes Pockels's contribution to science received recognition toward the end of her life. She was commemorated by an appreciative article written by Wilhelm Ostwald for her seventieth birthday, to which reference has already been made. About the same time the Carolina-Wilhemina University in her hometown of Braunschweig awarded her an honorary doctor's degree. That must have been quite a triumph for "Tante Agnes," who, as a girl, never went further than "Mädchenschule."

CHAPTER TWELVE

Comrades in the Search. The Flavor of Late Nineteenth-Century Physics

In chapter 3 short biographies of some of Benjamin Franklin's contemporaries were given, scientists with whom he associated in London when not engaged in diplomatic work. These biographies served to illustrate the flavor of the times, the enlightened amateur spirit. The present chapter provides similar information for a sampling of Lord Rayleigh's and Miss Pockels's contemporaries.

They are quite different from their eighteenth-century counterparts: science is their all-absorbing occupation and most of them are actually famous—giants in the history of science. Nevertheless, they are actually a remarkably varied group. They lived in different countries and had different kinds of careers. They were diverse in scientific methodology and style, and they had distinct personalities, ranging from gregarious to reclusive to suicidal.

Within the German-speaking countries there was less variety than elsewhere among scientific careers. Academic careers tended to follow an established pattern, one feature of which was that science professors moved frequently from one university to another as they advanced in rank or stature. The profession was usually entered under the patronage of an older professor; and an invitation to accompany one's patron on a move from one place to another was often the first sign of recognition for a younger assistant. Later, the patron would help the assistant obtain an independent post when some professor had died or a new position had been created. Then the younger man would himself start the cycle again, acquiring his own students and assistants, and so forth. This is a pattern that tends to generate a direct transfer of ideas and methods from one generation to the next—a master and disciple relation—and it is often easy to trace such sequential influences in the lives of German scientists.

There was less of this in Britain. It has been said of Lord Kelvin, for example, that he was no man's disciple and followed in no living scientist's footsteps. And in the United States there were as yet no footsteps to follow—Willard Gibbs had no predecessors at Yale.

Agnes Pockels, of course, did not know any of the contemporaries we list here and may not even have heard most of their names. She had no appointment and no colleagues in the normal sense, apart from her brother. Naturally, her name became known after her initial publications and Lord Rayleigh's evident commendation; and his sponsorship of her work must have come as something of a shock to at least some of her compatriots, accustomed as they were to orderly progress in personal advancement and achievement. Polite gestures were made toward Agnes Pockels at this point, offering her limited access to the academic community. Woldemar Voigt, regular professor of theoretical physics at Göttingen, director of her brother's doctoral thesis, and (like her brother) a crystal physicist, offered her the use of laboratory facilities at Göttingen's Physical Institute in 1893. Eduard Riecke, another physics professor at Göttingen, also interested in electrical properties of solids, is someone else she met at about the same time.

After Friedrich Pockels moved to Heidelberg in 1900, Agnes was actually invited by Georg Quincke to visit and discuss surface tension. Quincke was successor to the great Gustav Kirchhoff as professor at Heidelberg and was more prominent than Voigt or Riecke. He had entertained Lord Rayleigh in Heidelberg in 1883, and they corresponded about diffraction gratings and other subjects. Quincke also received honorary doctorates from both Cambridge and Oxford. However, by 1900 Agnes Pockels had little time left for research, because of the demands made on her by the need to care for her increasingly enfeebled parents.

Lord Rayleigh, on the other hand, knew all the people on our list (except perhaps Friedrich Pockels), either personally or through correspondence, or because their papers were related to his own work. And undoubtedly all of them knew who Lord Rayleigh was, and—except for Maxwell, who was no longer alive—they had read about and appreciated Rayleigh's measurement of a molecular dimension.

William Thomson (1824–1907). William Thomson, later Lord Kelvin, was a gregarious fellow and an international celebrity. He was more honored than any other physicist of his time, better appreciated than even Rayleigh or Maxwell, partly no doubt because of his public works and inventions—Kelvin did not limit his efforts to pure science.

Kelvin's father, James Thomson, was professor of mathematics at the University of Glasgow. He was intensely ambitious for his son. He guided him (even "pushed" him) into paths that would lead to achievement and fame, and he was instrumental in promoting William's appointment (at age twenty-two) to the post of professor of natural philosophy at the University of Glasgow. Kelvin held that position for an incredible term of fifty-four years, firmly rejecting efforts to lure him to Cambridge, the circumstances of which we have already related. Later, when he was made a peer, he chose the name "Kelvin," after a stream that runs through the university campus. (His title, of course, unlike Rayleigh's title, was not obtained by inheritance. It was conferred upon him by Queen Victoria in 1892, in recognition of his work and as a reward for the prestige that he brought to British science.)

Kelvin's success was fabulous. He published over six hundred papers and was granted seventy patents, many of them related to telegraphy and navigation. He is said to have been single-handedly responsible (against all kinds of cost-motivated pressures) for the success of one of the most ambitious and far-reaching projects of the late nineteenth century, the laying of the transatlantic cable between America and Europe. He also designed a new mariner's compass for use on iron ships as well as some other instruments, and he became wealthy from such commerce-related activities. Like many another successful self-made man (and quite unlike Lord Rayleigh), Kelvin took a turn to extravagance in his middle years. He bought a magnificent sailing ship in 1870 and entertained Helmholtz there a year later. After his second marriage in 1874 (age fifty), he built a baronial mansion at Largs, about twenty miles from Glasgow, and called it Netherhall. James Clerk Maxwell provided peacocks to grace its grounds.

Kelvin began his scientific career as an extraordinarily precocious applied mathematician with a highly individual style; he emphasized mathematical analogies with no necessary implication that they were manifestations of underlying physical analogies. One such analogy (between electricity and heat flow), which he published when only seventeen years old, was used by Kelvin's fellow Scotsman James Clerk Maxwell as a starting point for his theory of the electromagnetic field. Kelvin next turned to thermodynamics and the work that guaranteed his everlasting fame as a physicist—the recognition that the link between "temperature" and the irreversibility of energy transformations (the second law of thermodynamics) permits the definition of an *absolute scale of temperature*, independent of the properties of the substance used to measure it. This is the universal scientific temperature scale we use today, and it is appropriately called the Kelvin scale.

But another application of Kelvin's interest in the flow of energy, this time the flow from the sun to the earth, was much less successful. Kelvin estimated in 1854, using all available knowledge about heat sources and heat transfer, that the sun's age was then thirty-two thousand years, and he predicted that the sun's heat could last no longer than another three hundred thousand years. This may have been a reasonable conclusion on the basis of the laws of physics, but it was wildly unreasonable in the face of the geological record, and utterly irreconcilable with the theory of the evolution of living species, then being formulated (and soon to be published) by Charles Darwin. Kelvin later conceded that his calculations could be "stretched" to raise his estimates considerably, but he remained far short of the geological or biological age of the earth (around 4 billion years). Kelvin never became reconciled to the latter, even after natural radioactivity was discovered and provided the energy source that was needed to raise his previous estimates. In fact, he was unwilling to believe in the accepted interpretation of radioactivity. "The disintegration of the radium atom is wantonly nonsensical," he said.

This is by no means the only example of Kelvin's intransigent opposition to new ideas. His "pure" science became increasingly destructive as he got older. Since Kelvin and Rayleigh were good friends, the difference between them in this regard is noteworthy. We saw earlier that Rayleigh was himself not brilliantly innovative, and he never allowed the spurt of great discoveries in the 1890s to occupy much of his attention. But his was an attitude of benign neglect, whereas Kelvin was an active opponent who vigorously attacked almost all newly developing concepts.

Understandably, his antagonism toward new ideas has tarnished Kelvin's reputation. But there is no evidence that it did so while he was still alive. In 1902 (at age seventy-eight), on his fourth visit to the United States, Kelvin participated in the installation of Nicholas Murray Butler as President of Columbia University. Some students spotted him in the academic procession (he wasn't hard to spot with his shiny forehead and full white beard), and a shout went up, "Hats off to Kelvin." There are exceedingly few figures in the history of science who could have brought forth such a salute.

James Clerk Maxwell (1831–1879). James Clerk Maxwell, though lacking a title, was, like Lord Rayleigh, born into the landed gentry, and he never lacked the means for a comfortable life. He was actually not a "Maxwell" at all, but belonged to the Clerk family, from Penicuik near Edinburgh. His father, John Clerk, had to adopt the surname of Maxwell (for complex legal reasons) when he inherited an estate in "Maxwell territory," near Dumfries in southern Scotland, and moved his residence there. The estate, Glenlair, is where James Clerk Maxwell grew up, and it was to remain his home for most of his life.

Maxwell is rated by physicists themselves as one of their legendary figures, comparable to Isaac Newton and Albert Einstein. (In 1931, at the celebration of Maxwell's centennial, Albert Einstein called his contribution "the most fruitful that physics has experienced since the time of Newton.") His eminence was probably not quite as evident to his contemporaries, though Lord Rayleigh certainly recognized it. "I had the sense to cultivate Maxwell as much as I could," he said retrospectively, referring to his early years.

Maxwell studied at Cambridge, became a Fellow of Trinity College after he received his degree, but then returned to Scotland in 1856 as professor of natural philosophy in Marischal College, Aberdeen. From 1860 to 1865 he held a similar position at King's College, London, but in 1865 (having no real need for an outside income) he retired to his estate at Glenlair in order to devote full time to writing his celebrated *Treatise on Electricity and Magnetism*. He was recalled to academic life in 1871, as the first Cavendish Professor at Cambridge University, as has already been related. He was throughout his life a poor lecturer, lacking the charisma that Kelvin, for example, possessed, and he was a dismal failure in his academic positions. He was effectively "fired" from his job at Aberdeen, for example. In 1860 he applied for the chair of natural philosophy at the University of Edinburgh but was turned down in favor of his preparatory school classmate, P. G. Tait, who was certainly not his equal as scientist though he was a better teacher. Maxwell was a huge success at Cambridge, of course, but he had a different mandate there, the creation of the Cavendish Laboratory, a job that he did superbly.

Maxwell died of cancer at the early age of forty-eight, but, despite the brevity of his career, he contributed brilliantly to a remarkable variety of subjects: the composition of Saturn's rings, geometrical optics, an explanation of color vision, etc. He is especially famous for his work on the statistical molecular theory of heat and matter. He was the first scientist to introduce a probabilistic—i.e., basically indeterministic—factor into our view of matter, through the Maxwell distribution of velocities of gas molecules. This subject was subsequently developed more fully by Ludwig Boltzmann and Willard Gibbs. (See below.)

Maxwell's most brilliant achievement, however, was his theoretical work on electricity and magnetism. Building on the already known intimate relation between electricity and magnetism (established by Michael Faraday), he created a "complete" theory of the electromagnetic field, equations by which all conceivable manifestations of it could be expressed and understood. And this led to an early example of something that is more common today, true *pristine discovery from theory*, not experiment. It was one of the momentous scientific discoveries of all time, the discovery that ordinary light, the subject of so much controversy through the preceding centuries, was a form of electromagnetic radiation. The discovery came about because the velocity with which electromagnetic waves move through space (numerical value then unknown) could be determined from other known parameters by means of Maxwell's equations. The value obtained, 3×10^{10} cm/sec, proved to be the same as the measured speed of light. "Great guns," Maxwell called it in a letter, with uncharacteristic lack of modesty.

J. J. Thomson (1856–1940). J. J. Thomson was Rayleigh's immediate successor as Cavendish Professor at Cambridge and also followed him (in 1905) as professor of natural philosophy at the Royal Institution. The two men were undoubtedly in almost constant contact for about thirty years. The Cavendish Laboratory came to full flower during Thomson's directorship—seven winners of the Nobel Prize were trained there, for example. Thomson himself discovered the electron in 1897 and thereby gracefully led the laboratory through the transition into the new physics of subatomic particles. The contrasts among our four Britons, all trained in the old traditions of Cambridge mathematical physics, are noteworthy: Maxwell did not live to see the transition of old to new, Kelvin bitterly opposed it, Rayleigh treated it with benign neglect, and Thomson embraced it and encouraged his students to do the same.

Thomson differed from the others in another way: he was initially very poor. He was an engineering student at Owens College in his home city of Manchester, hoping to become apprenticed to a commercial engineer for practical training, but the death of his father made it impossible for the family to pay the normal apprenticeship fee. Thomas Barker, professor of mathematics at Owens, a former senior wrangler at Trinity College, Cambridge, fortunately took an interest in Thomson and advised him to try (with Barker's recommendation) for a Trinity College scholarship. Thomson succeeded, moved to Cambridge in 1876, finished second wrangler in the mathematics tripos, and then accepted a Trinity Fellowship. He was to remain in Cambridge for the rest of his life, and eventually (in 1918) he became Master of Trinity, one of the greatest honors that English science can bestow. He also won a Nobel Prize (in 1906) and was awarded a knighthood. Cambridge's notorious rule, forbidding collegiate fellows to marry, was abolished in 1882 and Thomson was able to have a normal family life outside his college. His son, George P. Thomson, also became a physicist and, in due course (1937) a Nobel laureate. J. J. Thomson had by then become a moderately wealthy man.

Thomson's discovery of the electron was an ideal instrument for ushering in the age of subatomic particles, for the experiment that led to the discovery was designed with no such result in mind. It arose out of a long-range experimental project on the electrical discharge through rarefied gases—the same "cathode rays" which Roentgen was studying when he discovered X rays. The nature of the rays was still controversial, though there was good evidence that they consisted of charged particles, most likely atoms with an excess of charge ("ions" in modern terminology). Thomson simply set out to measure the ratio of charge to mass by simple application of Newtonian mechanics. The astonishing result was that the charge/mass ratio was a thousand times larger than the same ratio for a proton, which is a hydrogen atom with a single charge on it. It could be considered virtually certain that the magnitude of the electric charge was the same for the two particles. Therefore, the mass of the cathode ray particle must be a thousand times smaller than the mass of the smallest known atom (compare table 3)!

Hermann Helmholtz (1821–1894). Hermann Helmholtz was another giant of nineteenth-century science, perhaps one of the last with a romantic yearning for grand unifying principles. He himself became famous for one of the greatest of all unifying principles, the first law of thermodynamics, the law of conservation of energy. This law grew gradually from diverse observations by many different people, and one cannot point to a single moment when one individual stood up and said, "I announce a new universal law of physics." But Helmholtz came close to filling that bill in 1847, with the publication of a creative work, *Ueber die Erhaltung der Kraft*.

Helmholtz's father was a schoolteacher and an ardent lover of philosophy. Young Hermann first wanted to become a physicist, but his father could not afford the fees for the requisite university education. He therefore took a degree in medicine instead (at the University of Berlin), for which there was a state subsidy. He became a pupil of the great physiology teacher, Johannes Müller, and did the thesis research required for an M.D. degree under his direction. In 1848 another former Müller student resigned as professor of physiology at the University of Königsberg to assume a similar post in Vienna, and Helmholtz was appointed associate professor in his place. In 1855 came another move, to the then vacant chair of anatomy and physiology at the University of Bonn, a position that Müller himself had held from 1830 to 1833. In 1857 the government of Baden, then a separate state, offered Helmholtz the chair of physiology at Heidelberg, which he accepted. He remained there for thirteen years, probably the most productive period of his career, and toward the end of this time he returned to his original love of physics. In 1871 the University of Berlin invited Helmholtz to become professor of physics. He accepted the post, lured by a high salary and the construction of a new institute of physics, which was to be entirely under his control. He soon became the acknowledged patriarch of German science, and acquired wealth and political power.

Helmholtz's specific contributions to science are numerous and extraordinarily varied. He always retained the interest in philosophy inspired by his father, especially in epistemology, in the question of how knowledge is acquired. He saw this problem as physical and studied acoustics (including the physics of music) and vision in part because he was thereby looking at the initial receptor for signals that ultimately became part of "knowledge." As a physiologist, Helmholtz was the first to make an experimental measurement of the rate of conduction of a signal along a nerve. He also invented the ophthalmoscope, the instrument still used today to obtain a magnified image of the retina of the human eye.

Helmholtz was still a physiologist in Müller's group in Berlin when he published *Ueber die Erhaltung der Kraft*. It was intended in part to support the group's conviction that life itself contributed no mystic "vital force" to the actions of living things—for example, they asserted, the huge force capacity of muscle movements can be entirely accounted for by the energy derived from food intake. But it was the recognition of the far-reaching importance of the concept per se that was the basis for Helmholtz's ultimate conversion to physics for its own sake.

Helmholtz's friendship with Kelvin and Rayleigh was unusually close, considering their different national allegiances. His friendship with Kelvin was personal as well as scientific—in 1871, for example, Helmholtz sailed the Scottish seas on Kelvin's yacht for a month. In the same year Helmholtz was offered the Cavendish Professorship at Cambridge, as already related. Ten years later Cambridge

University awarded Helmholtz an honorary degree, and he stayed with the Rayleighs at Terling for two nights. Rayleigh's son wrote later that he remembered the occasion very well, being greatly impressed when he was told that the visitor was an even cleverer man than his father.

Ludwig Boltzmann (1844–1906). Ludwig Boltzmann was born in Vienna and educated at Linz and Vienna. He was appointed mathematics professor at Vienna in 1873, and later held successive professorships in physics at Graz, Vienna, Munich, and Leipzig. He is famous for one of the great intellectual achievements of the period, the reconciliation of an apparent paradox between the molecular theory of matter and the empirical laws of thermodynamics. Most properties of gases can be explained by the "helter-skelter motion" of gas molecules (chapter 9), but the myriads of interparticle collisions in this motion, surely obeying Newtonian mechanics, must be individually *reversible*. How then do we account for the irreversibility of spontaneous events, as expressed by the second law of thermodynamics?

Boltzmann solved this problem (in a sustained effort spanning many years) by applying statistical analysis to molecular motion, as Maxwell had first done (see above). The brilliant innovation was to demonstrate the existence of physical laws, such as the second law of thermodynamics, that were properties of multitudes of molecules and not directly applicable to the behavior of individual particles. The very concept of such a thing was a profound novelty, in no way perceived by Newton or anyone else before the time of Maxwell and Boltzmann. The generation of this new concept involved the creation of a new fundamental constant of physics, called the Boltzmann constant, a kind of scaling factor between the energy units that we use in everyday life and the energies of molecular motions. This scaling factor (usually designated by the single letter k) is part of many formulas, including the "Boltzmann distribution law" and the famous equation for entropy:

$$\text{entropy} = k \log \boldsymbol{\Omega}$$

in which $\boldsymbol{\Omega}$ is a statistical factor, the number of different ways of distributing energy and position among multitudinous molecules without affecting observable bulk properties.

Success, alas, did not bring joy. Boltzmann became completely convinced that the molecular particles forming the basis of his theory must actually exist. In Britain that conclusion would have been considered sensible, perhaps not entirely proved but certainly acceptable. Many of Boltzmann's European contemporaries, however, were adamantly opposed to the idea and angered by Boltzmann's insistence. At conferences and elsewhere, over and over again, he was hounded by their skepticism. Isn't this just an abstract mathematical construct? Who has ever seen an atom?

Looking back, we would think that the discovery of radioactivity and of the electron would have been interpreted as "almost seeing" an atom and would have muted the opposition, but that didn't happen. (The Austrian philosopher Ernst Mach disputed the existence of atoms with Albert Einstein as late as 1912!) Boltzmann committed suicide in 1906 by hanging himself while on holiday near Trieste, then part of Austria. Persistent opposition to him was not his only motivation, but it cannot help but have contributed to his despondency.

Wilhelm Conrad Roentgen (1845–1923). Wilhelm Roentgen was not one of the giants of nineteenth-century physics, not in the same class with Boltzmann or Helmholtz, but his accidental discovery of X rays in 1895 was one of the grand spectacular events in the history of science. News of it spread almost instantly around the world, and the discovery was appreciated as heralding the dawn of a new age by scientist and layman alike.

Roentgen was the son of a cloth merchant from the Rhine province, and he had a typically mobile Germanic career. He earned a doctor's degree from the Polytechnic Institute in Zürich in 1869, became an assistant to a physics professor there, moved with him to the University of Würzburg in 1871, and shortly thereafter to Strasbourg, which had become German territory after the Franco-Prussian War, and where a new German university was being organized. Roentgen was given his own chair at the University of Giessen in 1879, and in 1888 he moved back to Würzburg as professor of physics and director of the physical institute. In 1894 he became rector of the university, an administrative post that appears not to have interrupted his research, because the discovery of X rays came the following year. In 1900 Roentgen became director of the physics institute in Munich, and he remained there until he retired in 1920.

Roentgen was a retiring and unassuming man. He was a meticulously conscientious experimenter who built most of his own apparatus and invariably worked alone in his laboratory to avoid distractions that might interfere with his acute powers of observation. He did not pursue a single research goal but worked at different times on diverse problems in the realm of physics, including electromagnetism, kinetic theory of gases, properties of crystals, etc. In personality he was therefore much like Lord Rayleigh, though he was less prolific and never attained the kind of eminence that Rayleigh did. Roentgen never attempted to explain the theoretical origin of X rays, and in fact returned to his study of the physics of solids after publication of his observations. This, too, was rather like Rayleigh. Roentgen was awarded a Nobel Prize for the discovery of X rays in 1901 (the very first such award to be made), and, just as Rayleigh would do later, did not keep the prize money, but donated it to the University of Würzburg.

Roentgen's single published paper on the spreading of oil on water was rather diffident. He had read Rayleigh's report, and it reminded him of experiments he had done several years before, with the same objective of measuring the thickness of an oil film on water. His method of observing the state of the surface involved blowing ether vapor onto the surface (without oil and then with increasing amounts of oil); and he reported two values for the thickness of a coherent film (being quite explicit in the use of the adjective "coherent"). One was 18×10^{-8} cm, the other was 5.6×10^{-8} cm, and it is not clear from his paper which of the two is supposed to represent a unimolecular film—in fact, Roentgen makes no explicit reference to molecules at all in his reminiscent report. Rayleigh does not mention Roentgen's paper in his later publications on the subject, but Pockels gives a reference to it in her second paper.

Friedrich Pockels (1865–1913). Friedrich Pockels was given the good schooling that his sister Agnes was denied; he attended the University of Göttingen, receiving a doctor's degree there in 1888. He was a competent theoretical physicist, but (in part perhaps because of less than robust health) never received a real professorial appointment. He had to be content with positions of "extraordinary" professor (positions created explicitly for individuals who were not selected for "regular" chairs), first at the Technische Hochschule in Dresden from 1896 to 1900, and then in Heidelberg from 1900 until his untimely death in 1913.

Nevertheless, Pockels's scientific achievements were notable—more lasting than those of some of his better-rewarded colleagues. While still an assistant at Göttingen, he published a book on the solutions and applications of a common differential equation. In the same period he made an important contribution to meteorology with a paper on the mathematics of cyclic motion of air masses. Most of his work, however, dealt with electrical, magnetic, and optical properties of crystals. He is one of the founders of "piezoelectricity," the phenomenon of the dependence of electrical conductivity in certain crystals upon pressure, and he remains well known today for his work on crystal optics. He has the honor of having an important phenomenon named after him, the "Pockels electro-optical effect," and his name is attached to an important device for electro-optical measurement, the "Pockels cell," developed in the 1940s on the basis of the effect he discovered. The Pockels cell has many uses: as a switching device in lasers, for example, and as a compensating device in instruments for the measurement of chemically induced "optical rotation," that is, change in the plane of polarization of light.

From 1908 until his death, Pockels was editor of the "Beiblätter," a review supplement to the *Annalen der Physik*. His former Göttingen mentor, Woldemar Voigt, wrote an obituary for Pockels in the "Beiblätter" after his death. He described Pockels as not getting much joy out of his professional life, because of poor health and disappointment at never receiving proper recognition. Pockels was married in 1900, and Voigt named his home life and his hobby of botanical studies as the sources of his greatest happiness.

Josiah Willard Gibbs (1839–1903). Josiah Willard Gibbs was one of America's earliest academic scientists, and some contend that he is still the best we have ever had, notwithstanding our plethora of modern superstars. His work was done in virtual isolation, there being no one of comparable interests or ability with whom he could discuss it.

Gibbs was the son of a professor of sacred literature at Yale University's divinity school. He did both his undergraduate and graduate studies at Yale, receiving a Ph.D. degree in 1863, which was only the second science Ph.D. ever awarded in the United States. The year 1863 was the year of the Battle of Gettysburg in the American Civil War, but Gibbs was ineligible for military service because of a suspected tendency to tuberculosis. After the war was over, he spent three years studying in Paris, Berlin, and Heidelberg. While there, he attended lectures and studied and absorbed their contents, but there is no record of significant personal interactions with his professors or even with other students. Gibbs tended to remain an anonymous figure in the back of the crowd throughout his career. He almost never attended scientific meetings—even when awarded the Rumford Medal by the American Academy of Arts and Sciences in 1881, he declined to make the short trip to Boston to receive the honor in person.

Gibbs was appointed professor of mathematical physics at Yale in 1871—the year of Maxwell's appointment to the Cavendish chair in Cambridge. For the first ten years, during which he did his most important scholarly work, he received no salary. In fact, all that Yale gave him for his services was a place to work in, and that was no more than a small study in one of the student dormitories. (It is not clear whether this was because Gibbs wanted to be an amateur, or because Yale University didn't think science education had enough value to merit a paid position.) Gibbs accepted an invitation in 1879 to give a series of lectures at the then newly founded Johns Hopkins University in Baltimore, and shortly thereafter John Hopkins offered him a professorship at the princely salary of $3000 a year. Gibbs was inclined to accept, but Yale shrewdly made a counteroffer of $2000 a year, guessing (correctly) that Gibbs would not want to move away from New Haven for a mere difference of $1000 a year. Gibbs remained at Yale for the rest of his life. He never married, but lived with his sisters as long as they remained alive, and with the family of one of them after that.

Scientifically, Gibbs was quite unlike most of the famous European physicists. He was not a polymath. He worked on just a small number of projects, but did so with extraordinary thoroughness and concentration, so that what emerged at the end was a "finished" piece. His masterpiece was a 320-page paper entitled "On the Equilibrium of Heterogeneous Substances," published in two parts in the *Transactions of the Connecticut Academy* between 1875 and 1878. What the paper does is to transform the inexorable laws of thermodynamics (conservation of energy and the *irreversible* element in its utilization) into an entirely new discipline—a tool whereby the direction of change for virtually all real phenomena can be understood and whereby the equilibrium states, the states where all change stops, can be confidently predicted. The marvel is the ability to bring highly *specific* phenomena into the realm of simple *universal* physical law. Gibbsian thermodynamics applies to mixtures as well as pure substances, to surfaces as well as bulk phases (see

Langmuir's application on p. 172, for example), to complex chemical reactions, and even to the life and death of biological cells.

Gibbs understood the revolution he had wrought. He was not boastful about it, but he understood physics and the important place that his paper would inevitably hold in the framework of that science. He knew also that the *Transactions of the Connecticut Academy* were not widely read and he sent about ninety reprints of his papers to people in America and Europe who he assumed might be interested, including Maxwell, Rayleigh, Boltzmann, Kelvin, Clausius, etc. To some of them he even sent copies of printer's proofs before actual publication. His work was gradually recognized at its true value, though Kelvin, as usual, was skeptical. "I find no light or leading for either chemistry or thermodynamics in Willard Gibbs," he wrote in a letter to Rayleigh as late as 1892. Rayleigh himself was one of Gibbs's admirers, and was instrumental in the selection of Gibbs as recipient of the Royal Society's Copley Medal in 1901. Boltzmann sent Gibbs a personal invitation to attend a meeting in Germany, but Gibbs declined.

Henry Rowland (1848–1901). Henry Rowland was perhaps more representative of American physics than Willard Gibbs. He was the son of a prosperous clergyman and was himself intended by his family for the ministry. But he developed an interest in chemistry and electricity and engineering, and resolved to find a career in science. He began in 1872 as an instructor in physics at Rensselaer Polytechnic Institute in Troy, New York, and in 1875 he was chosen as the first professor of physics at Johns Hopkins University in Baltimore, then newly founded as an ambitiously conceived experimental institution, emphasizing graduate research rather than undergraduate collegiate teaching. He remained at Johns Hopkins for the rest of his life.

Rowland, like many academic Americans then and now, was a "builder"—a promoter of science, a creator of organizations, someone who tried to recognize other scholars (and to hire them) rather than being one himself. When first appointed to the Johns

Hopkins chair he went to Europe for a year, not to learn physics (as Gibbs had done ten years earlier), but to inspect laboratories and to purchase equipment; he returned home in 1876 to make the Johns Hopkins physics laboratory the first effective teaching laboratory (certainly the best equipped) in the United States. He spent four months in Helmholtz's new laboratory in Berlin during this trip and lesser amounts of time with other European physicists. On the whole, he was more impressed by the laboratories than by the professors, but Maxwell was an exception: "After seeing Maxwell," he said, "I felt somewhat discouraged for here I met with a mind whose superiority was almost oppressive." His judgment was evidently good.

While at Johns Hopkins, Rowland was the catalyst for inviting Willard Gibbs to that university. (See above.) When the British Association for the Advancement of Science held an unprecedented overseas meeting in Montreal in 1884, Rowland persuaded Lord Kelvin to come to Johns Hopkins after the meeting to deliver a series of subsequently famous lectures, the "Baltimore Lectures," which have recently been reprinted. Rayleigh met Rowland on this occasion and wrote home about it, calling him "about the first physicist in America." (Gibbs was apparently temporarily forgotten. He attended the Montreal meeting, but, as usual, remained in the background.) In 1899 Rowland became one of the founders of the American Physical Society and was elected to be its first president.

Rowland's own research was on a very modest scale. He determined the definitive value for the electrical unit of resistance, the ohm, and invented a device for making better diffraction gratings for spectroscopy than anybody else could make. He manufactured and sold about one hundred of the gratings throughout the world, selling them at cost.

CHAPTER THIRTEEN

Ben Franklin Wonders Why (Molecular Interpretation)

We take off in a completely new direction now. We are no longer asking whether atoms and molecules exist or groping to understand those who opposed the concept. We are no longer asking about molecular size. We know how small a molecule is.

The new direction is to ask about molecular *behavior*. After all, the spontaneous spreading of the oil is truly unexpected, even through our own modern eyes, accustomed to marvels of nature and technology. Knowing about molecules and their size makes it, at least numerically, even more startling. One teaspoonful of oil spreads to half an acre; that constitutes a 10^7-fold increase in area. Translated into molecular terms it becomes a layer that is everywhere one molecule thick and 10^{21} molecules in cross section! What properties of the molecules are responsible? Is it possible to fit this extraordinary result into the framework of established physical law?

A fascinating aspect of this problem is that Lord Rayleigh, who knew about molecules and their sizes, who had himself done the spreading experiment deliberately "for the determination of molecular magnitudes," never asked the question. He never wondered why the oil molecules do this extraordinary thing of spreading themselves out until there is but a single molecule between the water and the air above it. Or, if he did wonder, he kept it to himself.

Whereas Benjamin Franklin, who had at best a superficial notion of a particulate universe and could not have known about atoms and molecules in the modern sense at all, wondered why from the moment he first observed the phenomenon.

"I think it a curious enquiry, and I wish to understand whence it arises," he wrote at one point, and, a little later, "This prevention (of wave formation) I would thus endeavour to explain." What follows in Franklin's text makes it evident that when he used the words "understand" and "explain" he meant understand and explain in terms of forces between molecules. His "enquiry," though unsuccessful and perhaps even causal—there is no evidence that he spent much effort on it—is in language that we understand today. It is as if he could visualize the molecules of matter in his mind.

Franklin's "endeavour to explain" was unsuccessful, not only to modern eyes but probably in his own mind as well, for the explanation in Franklin's paper lacks the enthusiasm and verbal precision that characterize the description. The failure to explain can be excused in part by saying that the time was not ripe for a correct molecular interpretation—but only in part, for we can easily recognize a flaw in the logic of Franklin's speculations that he should have spotted himself. He could have done better than he did, if he had given more thought to the problem. Perhaps his mind was burdened by weightier matters.

A few direct quotations follow to illustrate Franklin's train of thought:

> If a drop of oil is put on a polished marble table, or on a looking glass that lies horizontally, the drop remains in its place.

And he might have added, but did not, that a drop of *water* on a marble table or on a looking glass also does not spread, a fact which he must have known and presumably had in mind. Everyone knows that drops of water don't spread out. It doesn't need saying.

In any case, it is clear that both oil and water molecules must somehow play a part in the explanation of the spreading phenomenon, and Franklin also invokes air molecules, because air molecules, in the form of wind, raise the waves.

> There seems to be no natural repulsion between water and air, such as to keep them from coming into contact with each other. Hence we find a quantity of air in water; and if we extract it by means of the air-pump, the same water again exposed to the air, will soon imbibe an equal quantity.

As description, this is beautifully lucid and succinct. But it contains the uncomfortable word "repulsion." Is it safe to treat "no repulsion" as a double negative, to assume that Franklin imagined that there were actual forces of "attraction" between air and water molecules? Perhaps so, because a little later he uses "no attraction" to explain why oil (unlike air) does not dissolve in the water.

Unfortunately, after this, Franklin's reasoning goes awry. This time he uses "repulsion" in a positive sense, and there is no ambiguity in the meaning—Franklin was imagining a real force that kept molecules apart:

> But if there be a mutual repulsion between the particles of oil, . . . oil dropt on water will not be held together by adhesion to the spot whereon it falls; it will not be imbibed by the water; it will be at liberty to expand itself; and it will spread . . . till the mutual repulsion between the particles of the oil is weakened and reduced to nothing by their distance.

The trouble with this statement, however, is not that it invokes a repulsive force. The argument is internally inconsistent quite independently of one's view of intermolecular forces, for two reasons. (1) Franklin, just a few pages earlier, said that a drop of oil put on a polished marble table or a horizontal looking glass "remains in its place, spreading very little." Why should the oil particles (i.e., molecules) repel one another in one situation and not in the other? (2) Why, all of a sudden, does Franklin leave water molecules out of the equation, when he has up to this point taken it for granted that oil, water, and even air must all be involved in the phenomenon?

The correct explanation was of course not easy to arrive at, simple as it may seem to us now. It involves a paradox, the resolution of which requires a sophisticated view of the nature of molecules beyond the ability of anyone in the eighteenth century.

Figure 17 explains the paradox, but let us first return to figure 10 in chapter 8, which portrays the formation of a thin film as the

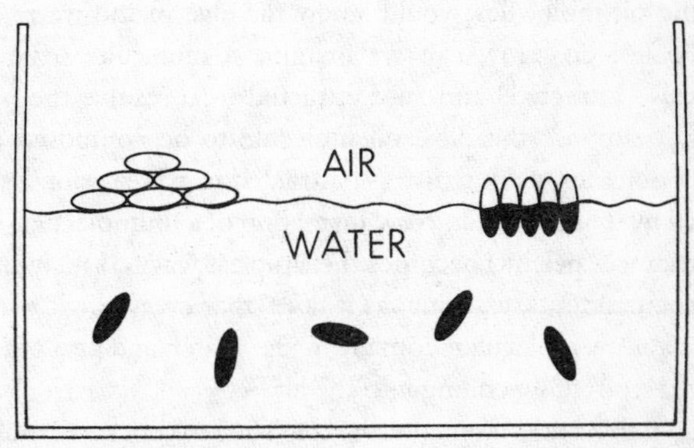

Figure 17. Formation of a monolayer. In this illustration white areas represent molecular surfaces *with no attraction for water*, and black areas represent molecular surfaces *with a strong attraction for water*. Totally "white" molecules do not enter the water at all. Totally "black" molecules are totally immersed in water. Molecules composed of two domains, attracted to water at one end and repelled by it at the other, form a monolayer—a film that is one molecule thick. That is the obvious way to satisfy the dual force requirements, assuring, as it does, that *every single molecule* will have one molecular domain in the water and the other one removed from it, extending into the air.

result of the tumbling of molecules from an initially compact drop. It explains why a unimolecular layer is the likely result once we imagine the existence of some force that creates tumbling. We ask questions in the figure caption. When will the process stop? Why should it stop at all until there are no more particles to tumble? But there is another question we do not ask—why should the molecules be driven to tumble from the drop in the first place?

The only possible answer is that oil molecules, normally comfortable when clinging to each other in a little drop, want to do something different when water is nearby: that they are attracted to water, seek contact with water.

The paradox is that we have already concluded that there can be "no attraction" between oil and water, for if there were attraction, then the oil molecules would surely dissolve in the water, which they do not do. How can we imagine a molecule to be simultaneously "attracted" and "not attracted"? To resolve the paradox we must suppose the olive oil molecule to be composed of two distinct domains, as in figure 17, attracted to water at one end but repelled by it at the other. Now formation of a unimolecular film (if circumstances permit) becomes the simplest way to satisfy the dual force requirements, assuring, as it does, that every single molecule will have one molecular domain in the water and the other one removed from it, extending into the air.

In 1770 this idea was inconceivable. The concept of a "molecule" as the smallest particle with the chemical attributes of a substance was widely accepted, though different terminology was used. But it was natural to assume that these defining attributes would be distributed uniformly throughout the mass of the molecule, that every part of each molecule would possess the same properties as the bulk substance from which it was derived. The idea that dramatic differentiation could exist within a single molecule could not have been conceived.

But that idea is precisely what is *essential* for an explanation of the spreading phenomenon. It requires the simultaneous presence of both attractive and repulsive forces, the attractive force to make sure that each oil molecule is at least in part immersed in the water, the repulsive force to make sure that at least a part of each molecule remains outside the water. Only in this way can a film one molecule thick be formed.

The following chapter will endeavor to provide an up-to-date explanation for this intramolecular dichotomy, on the basis of the general laws of physics and the quite special chemistry of water.

CHAPTER FOURTEEN
In Praise of Water

1. "Water is the Origin of All Things"

IN the science fiction novel *Dune*, author Frank Herbert takes us to Arrakis, imaginary third planet of the pilot star Canopus. Arrakis has an atmosphere much like that of Earth, but almost no water, and Frank Herbert creates for it an astonishingly believable human culture, shaped by the ever-present need for water. There are plastic still suits to prevent the loss of even minute amounts of water from perspiration, nose filters to recapture the moisture from exhaled breath, dew precipitators to harvest whatever moisture may have condensed at night from the arid air. And there is a recurring ceremony, the awesome ritual of the taking of the waters from the dead. *Dune* is considered a masterpiece of science fiction, because of its realistic portrayal of human character on the dry planet: valor, loyalty, greed, and cruelty are all sharpened rather than diminished by the harshness of the environment. But *Dune* is also a tribute to water, the common chemical all around us, which we take for granted until it is in short supply.

Turning from fiction to fact, we come to Thales of Miletus, living about 600 B.C. Some say he was the first scientist, the first in history to seek a natural origin for the behavior of the universe and the properties of matter. He made water the cornerstone. "The Earth floats on water," he said; and "Water is the origin of all things."

Thales was, of course, overenthusiastic. Water is not the universal prime mover. On the contrary, in an age of adulation for the marvels of chemistry, the chemical properties of water strike us as unspectacular. Water is noncombustible, neither acid nor alkaline, not a potent oxidant. Water is not used at all, even deliberately excluded, in many chemical or pharmaceutical manufacturing processes.

And yet, despite its apparent meekness, water merits some of Thales' praise. Potent destructive power is not the only way to dominance, and water, by more subtle means, can often exert a commanding influence on its environment. This is especially true for living systems, where smoke and flames and other flamboyant chemical reactivities would be destructive and cannot be tolerated.

Chemists, about a hundred years ago, recognized the importance of water and invented special terms to describe the behavior of other molecules with respect to water: *hydrophilic*, for molecules with high affinity for water, and *hydrophobic*, for molecules that appear to avoid contact with water. These terms were used irrespective of what other properties a molecule might have.

Applied to *entire molecules*, the words have little value, and in that context would not have survived long. They would be merely synonyms for high or low *solubility* in water, and one hardly needs special terms for that. In fact, there are today standard handbooks on every laboratory shelf that tabulate the solubilities of thousands of substances in water and in other solvents. We find not only very high or very low solubilities, but all gradations in between. The numbers are much more useful than descriptive words.

But we know now that many complex molecules—those belonging to the realm of organic chemistry—often contain particular groupings of atoms that retain their characteristic chemical properties when incorporated into larger molecules. They were called *radicals* by Liebig (see chapter 9). A more common term today is *domain*. Applied to domains rather than entire molecules, the words "hydrophilic" and "hydrophobic" make very good sense. A hydrophilic domain, if it were a molecule all by itself, would be very soluble in water. A hydrophobic domain, if it were a molecule all by itself, would have extremely low solubility in water. A hydrophilic domain, not by itself, but situated on a larger structure, retains the *tendency* to enter water. A hydrophobic domain, when part of a larger structure, retains the *tendency* to stay out of water. Thus hydrophilic and hydrophobic, applied to domains, express something more important than solubility, namely an assertion of individuality in different parts of a single molecule that would be difficult to express as clearly in other words.

The last paragraph of course raises the possibility that hydrophilic and hydrophobic domains may be present on the *same* molecule, exactly the sort of dual personality we were looking for in the preceding chapter.

But to assess the actual likelihood of this pivotal idea we must first provide some background information about molecules in the liquid state, using concepts that have to be derived from relatively modern physics. Even Irving Langmuir, who, as we shall see in the following chapter, had a brilliant ability to envision molecular attitude and shape, could not possess this theoretical background, so that in a sense we are getting ahead of our overall chronology. But that is quite appropriate, because Irving Langmuir did not himself ask the question, "why?", any more than Lord Rayleigh had done.

2. Atoms, Molecules, and Ions

Here I present just a little modern chemistry, just enough to explain water and how it interacts with other substances. The most important part is to understand how electrical charge is distributed within molecules.

Benjamin Franklin was one of the first to understand that "electric fire is a common element" of all matter, and he assumed it to be particulate. Michael Faraday in 1834 was the first to understand that electric particles of both positive and negative charge could exist in water solution to make such a solution electrically conducting. He called these particles "ions," after the Greek word for "going." But it was only after the discovery of electrons and other subunits *within the atom* (beginning in the 1890s) that electrically charged particles could be given a simple easy-to-understand explanation within the same framework that is used (now) to describe all of matter.

We now know that atoms are not structureless hard spheres, like the billiard balls that were used by nineteenth-century English theoreticians as models for colliding atoms or molecules. Atoms resemble billiard balls in that they are electrically neutral and spherically symmetrical, but the electric charge within them is quite specifically organized. There is a dense positively charged nucleus at the center of the atom, with a diameter of only about 10^{-12} cm. Negatively charged electrons occupy shells around the nucleus, defining the outer perimeter, which, as we have noted before, lies more than 10^{-8} cm from the atomic center.

The outermost electrons, relatively far from the positively charged nucleus, are relatively easily lost from an atom, leaving it with an excess positive charge—that is, transforming it into an ion. To be more precise, an electron cannot really be "lost." It must go somewhere, and in a liquid medium, where the density of surrounding matter is high, it is actually transferred to another initially neutral atom, which thereby acquires an excess negative charge. In the context of this book, therefore, which is not concerned with outer space or high vacuum laboratory apparatus, both positive and negative ions are normally present together. Superscripts "+" and "–" are used to designate the kind of charge. More than one electron can be gained or lost per atom, so that there can be more than one charge per ion. Sodium (Na^+), chloride (Cl^-), and calcium (Ca^{++}) ions are examples. In the normal earthly environment ions *usually* exist in ratios that exactly balance the total positive and negative charge. A crystal of common salt, for example, has overall neutrality though the building blocks of the crystal structure are Na^+ and Cl^- ions.

Molecules as well as single atoms have the ability to lose or gain an electron or two. The molecules are then called molecular ions. Bicarbonate ion (HCO_3^-) and ammonium ion (NH_4^+) are common examples. (Refer back to figure 11 in chapter 9, where the existence of molecular ions was already mentioned.)

Although ions are common everywhere, especially in water solution, they are still outnumbered in the world around us (typically by about 100:1) by electrically neutral molecules. The molecules in figure 11 other than ammonium cyanate are electrically neutral, and so is the olive oil molecule shown in figure 12. However, even such neutral molecules are *not uniformly neutral*. On the contrary, they tend to have an unsymmetrical distribution of electric charge within them, a little excess of positive charge at one or more places in the molecule and a complementary excess of negative charge at other places.

How does this come about? How does a nonuniform charge distribution arise in molecules known to be constructed from electrically symmetrical atoms? It is really quite simple. Atomic nuclei are not altered when molecules are formed and they remain the centers of positive charge within a molecule. But peripheral electrons—as in ion formation, the electrons farthest from the nuclei are involved—do move when molecules are formed. In fact, these electrons are intimate participants in the process of bonding atoms together, and many of them end up being shared between adjacent atomic nuclei instead of "belonging" to a single nucleus. Because individual atomic nuclei tend to have unequal intrinsic affinities for their outermost electrons, the "sharing" of electrons tends to be unequal too. The negatively charged electrons come closer to some nuclei than to others, and that is how unequal charge distribution is achieved. Neutral molecules with such unequal distribution of electric charge are said to be *polar*.

The water molecule is one of the best examples. In fact, there are very few other molecules that have so strong a polarity concentrated into so small a space. Figure 18 gives a structural diagram. The two constituent hydrogen atoms have *positive* polarity and the oxygen atom has *negative* polarity. This doesn't mean that the molecule is internally ionic, with a whole electron removed from

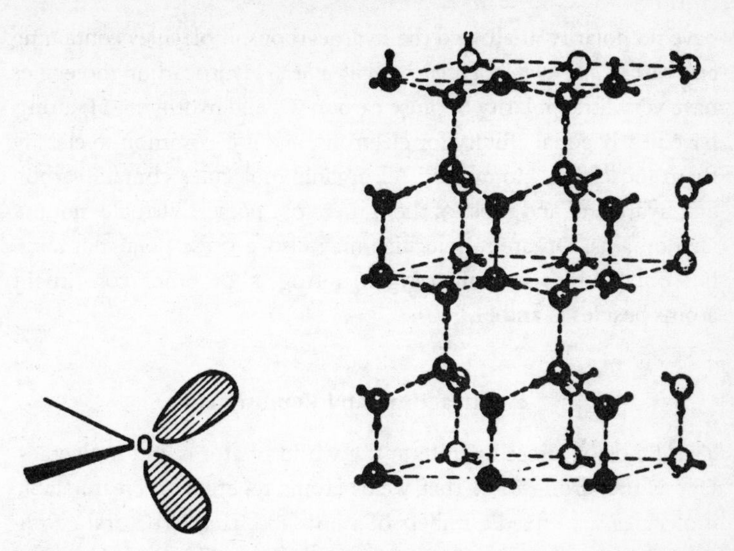

Figure 18. A water molecule is shown on the left. The two "bonds" to the left of the oxygen atom (one coming out of the page toward us, the other receding backward from the page) point to the two H atoms of the molecule. They represent poles of positive charge. There are two complementary poles of negative charge, both lying in the plane of the page: they are more diffuse and are shown in the picture as shaded areas. The hydrogen atoms out of the plane of the paper and the negative areas in the plane define what is meant by a tetrahedral arrangement. The structure of ice that results from the tetrahedral arrangement of positive and negative polarities is shown on the right.

each hydrogen atom—it would be incorrect to write $(H^+)(O^{--})(H^+)$. But there is enough asymmetry in charge distribution so that dissolved Na^+, for example, would be attracted spontaneously to the oxygen end of a water molecule, and dissolved Cl^- would preferentially associate with the other side.

At the opposite extreme to water are the inert gas "molecules," actually single atoms (helium, neon, argon, for example) which have no polarity at all, and the hydrocarbons, molecules containing *only carbon and hydrogen atoms and no others*. Hydrocarbon molecules have very little polarity because carbon (C) and hydrogen (H) atoms have nearly equal affinity for electrons, and it is common to classify them loosely as "nonpolar." All organic molecules contain carbon and hydrogen, and most of them therefore possess virtually nonpolar domains. Organic molecules may also possess polar domains, but only if they contain oxygen, nitrogen, or other constituent atoms besides C and H.*

3. Attractions and Repulsions

Two kinds of forces operate in the world of atoms and molecules. One is the strong force that welds atoms together to create stable molecules with fixed numbers of atoms and fixed structural geometry. Quantum mechanics is necessary to describe this force—the sharing of electrons that creates the bonds between atomic nuclei. This is one of the situations where classical Newtonian mechanics breaks down.

The second, weaker force is the force *between molecules* that leads to the existence of liquid and solid states. Without this weaker force, molecules would exist only in a gaseous state, each molecule moving about in space independently of the others. This weaker force, too, can involve quantum-mechanical elements, but a major component of it is often purely classical, an electrical force, plus

* Organic molecules, if they are large enough, can even have ionic domains, with a full unit of positive or negative charge per domain. Biological macromolecules, thousands of times larger than a water molecule, are so big that they can have many ionic domains per molecule, some positive and some negative, plus neutral polar domains plus non-polar hydrocarbon domains. Protein molecules, universal catalysts of biological chemistry, fall into this category. So does the DNA molecule shown earlier in figure 13. It actually has a full negative charge on each separate nucleotide building block, and that is what confers aqueous solubility on the molecule.

attracts minus, intuitively easy to understand. It is especially easy to understand for liquid water, as can be seen from figure 18. The figure shows that a single water molecule has its positive and negative "poles" in a tetrahedral arrangement, and the electrical +/− attraction extends as a result in all directions in three-dimensional space, and the packing of water molecules together follows the tetrahedral lines of force. In ice crystals H_2O molecules fill all space in a rigid array, as shown in the right-hand part of figure 18. In the liquid (fluid) state there can by definition be no long-range rigid structure, but the molecular polarity is the same, and tetrahedrally arranged clusters still dominate the molecular organization. The clusters are just much smaller than in the solid and able to move relative to each other. In addition, individual water molecules at cluster boundaries are also capable of motion, attaching themselves to one cluster at one instant and to an adjacent one the next.

What about other molecules (or ions) dissolved in water? The scenario of the preceding paragraph can be extended to them without introducing new problems. Ions, bearing whole electric charges, are usually very soluble in water. The force between a whole charge and the fractional charge at the corner of a water molecule is stronger than the force between the poles of adjacent water molecules, so that ions tend to be "hydrated," surrounded by water molecules that have been pulled away from clusters of pure water. Polar molecules with overall neutrality also tend to be soluble in water, their localized partial positive and negative charges able to attract water molecules and thereby to blend into the overall bulk clustered fluid.

Nonpolar molecules, according to this line of reasoning, should have little or no solubility in water, because they are unable to contribute an electrical interaction component to intermolecular attraction. Even when other (quantum-chemical) kinds of forces are taken into consideration, nonpolar molecules still tend to be excluded because they occupy space in the bulk liquid. Water molecules at the boundary of this space will tend to have one potential polar bond unsatisfied or, if they manage to make the bond to another water molecule by distortion of the usual tetrahedral arrangement, they will be unable to engage in their normal cluster dance, skipping from one cluster to another and back again. There is no actual active repulsion between individual nonpolar and water molecules, but the effect is the same. Water molecules move more freely if they have no nonpolar entities adjacent, and the total effect of the molecular dance, over a period of time, is to squeeze the nonpolar entities out. (Freedom of motion is part of "entropy." That is why one often reads that hydrophobic repulsion is an "entropic" phenomenon.)

4. Why Water Is Special

Water molecules are highly polar and exceptionally strongly attracted to one another. Water molecules are also very small, which gives them greater freedom to rotate or twist around than one might otherwise expect for molecules subject to such strong forces.

Small molecular size and high mobility enable water to be remarkably flexible in accommodating ions and other polar molecules. Several water molecules are needed to "hydrate" even a small ion. Individual water molecules can squirm around in many different ways to take advantage of the possibility of forming stronger bonds. And, likewise, the hydrophobic interaction with nonpolar entities is not limited to large areas or particular shapes for the interacting surface.

The size factor is vital here, for it makes water molecules smaller than the molecules of any solute molecule in which we are likely to be interested, smaller even than *individual domains* that may be part of a solute molecule. In defining the terms hydrophobic and hydrophilic near the beginning of this chapter, we asked what would happen if hydrophilic and hydrophobic domains were present on the *same* molecule. If water molecules had about the same size as the domains, then a competitive situation would arise, with the outcome difficult to predict. But the sizes are not the same. Water can recognize even small parts of molecules as attractive or antagonistic and can respond appropriately.

Looking back at the olive oil (triolein) molecule in figure 12, we see that the oxygen atoms at the base of the figure constitute a *polar* domain, whereas the three hydrocarbon chains extending away from it are hydrophobic. The polar domain will want to be anchored in the water, and the hydrocarbon chains will be uncompromisingly excluded. The dual personality of the olive oil molecule and the special character of water molecules make *possible* what at the end of the preceding chapter we characterized as *necessary* for the spontaneous formation of a film one molecule thick.

CHAPTER FIFTEEN

Irving Langmuir: Molecular Attitudes

Oleic acid on water forms a film one molecule deep, in which the hydrocarbon chains stand vertically on the water surface with the COOH groups in contact with the water.

Irving Langmuir, 1917

1. America on the Move

IRVING Langmuir was born in Brooklyn, New York, in 1881 and died on a visit to his nephew David on Cape Cod in 1957. His life span coincides with America's great years of expansion, her meteoric rise to the ranks of the world's top nations. America was a new force in the world, a force for progress and democracy and nationwide prosperity. Driven by boundless energy, supremely self-confident, it rode roughshod over traditional methods and social customs, sometimes forgetting that anything at all had existed before, and sometimes undoubtedly throwing out the good with the bad.

America encouraged personal ambition, which was "bad form" in some more traditional Western countries. Americans had an inexorable drive to get to the top, enjoyed the challenge of contests against the rigors of nature as well as in the world of business. "I wish to preach . . . the doctrine of the strenuous life," President Theodore Roosevelt said in 1899. (See Fig. 19.)

America stood for vigorous marketing and salesmanship. Making money was the name of the game, but with a new socially beneficial twist—the realization that bigger profits could come from mass consumption than from selling exclusively to the elite. Traditional class barriers were eroded to create more customers. Former emigrants—whose fathers or grandfathers had left Europe to seek their fortune in the New World of Opportunity—returned now as instruments of American industry. It was colonization in reverse, seeking new markets, looking for economic domination instead of territorial acquisition.

It was this drive for European markets that caused Irving Langmuir to be in Paris from age eleven to fourteen. His father was an insurance executive, head of the European agencies of the New York Life Insurance Company, directing sales throughout the continent from his headquarters in the French capital. The family lived in Paris, but traveled extensively to other parts of France, to Switzerland, to England, and to the Channel Islands. Irving loved Paris, especially the opera—he and his brother Herbert rarely missed a performance, though limited by their allowances to standing room in the uppermost gallery. But Irving didn't love French schools. They had rigid curricula, which stifled individuality. The pupils were supposed to absorb knowledge in silence, without asking foolish questions. His father decided to send Irving back home in 1896, to a private boarding school in Philadelphia where there was more encouragement of Yankee individual initiative.

Above all, America was attuned to constant change. Today was temporary, yesterday was obsolete, what mattered was tomorrow. Irving Langmuir sensed this as a child in 1896, when he returned home after his three years in Paris. "Trolleys everywhere and lots of electric lights and the tall buildings in New York all so changed in three years," he wrote in a letter. He could not know then, of course, that he was himself destined to help spread electric lighting to all corners of the world.

It is fascinating how the well-worn clichés that describe America in general apply to Irving Langmuir in particular. Self-confident, energetic, ambitious, an enthusiastic disciple of Roosevelt's prescription for a strenuous life, not only in his professional pursuits

but in his private life as well. He was an avid mountain climber and skier. He learned to fly and became a personal friend of Charles Lindbergh. While carrying out his graduate studies for the Ph.D. degree in Göttingen, he was challenged by a Dr. Jahn to a fifty-mile walk to the "Brocken." He describes it in his diary: "Got up at 1 a.m., cooked breakfast and started at 2.20. We walked 27 miles without stopping in 7½ hours."

In 1921, past forty and without preliminary conditioning, he climbed the Matterhorn: "We left at 2.40 a.m. Beautiful night with thin crescent moon. I found it hard climbing so fast in the dark, soon began to be winded easily . . . Suffered greatly from lack of breath and once or twice from faintness and sudden dizziness. Thought at one time that I could not reach top." But of course he did, arriving at the summit at 8.20 a.m. (Just as he did in his profession—aiming for the top and always getting there.)

2. General Electric

Giant corporations were an integral part of this ascending American society: Standard Oil, American Tobacco, General Electric, U.S. Steel. A little later, after Henry Ford introduced his mass-produced Model T in 1908, the great automobile companies joined the ranks. And much later, in Langmuir's last years, we all almost believed it when we were told that what was good for General Motors was good for the country.

Among these companies, General Electric, formed in 1892 from

Figure 19. Irving Langmuir in pursuit of the strenuous life. This picture was taken in the Adirondack Mountains by Vincent Schaefer, originally an uneducated apprentice machinist at the General Electric Company who later became Langmuir's close personal friend and his collaborator in the laboratory. The photograph is from the General Electric Company files and was obtained through the kindness of Dr. George Gaines at the General Electric Company.

successive mergers of smaller "Edison" companies, was unique. It was probably the first major corporation to recognize the need for a research laboratory—as distinct from (and in addition to) a pilot plant directly geared to production. General Electric's research laboratory employed academically trained scientists and encouraged them to do "pure" research, not necessarily applicable to a currently manufactured product but laying long-range foundations for possible future enterprise.

General Electric gave young Langmuir a job in 1909, to do whatever he wanted in the laboratory, with only the most casual review by the director of research. This is not quite as altruistic as it sounds, because the director knew what Langmuir wanted to do and that it was directly related to General Electric's biggest manufacturing headache at the time. Langmuir's doctoral research at Göttingen (under subsequent Nobel laureate Walther Nernst) had consisted of a study of chemical processes in gases in the neighborhood of hot incandescent filaments. He welcomed the opportunity to continue the work with the *much hotter* tungsten filaments that General Electric was then developing for its electric light bulbs—filaments that were still the source of exasperating practical problems. So Langmuir's work was unquestionably in General Electric's own best interests. Nevertheless, it is true that they granted him unparalleled freedom and that they waited patiently for several years without return on their investment.

Eventually the investment paid off, for Langmuir solved their problem. In order to do so, however, he needed to upset cherished dogma (the better the vacuum, the better the bulb), which is something that a researcher is not likely to do without the kind of freedom that Irving Langmuir was given. The ultimate result was a famous patent (applied for in 1913) for nitrogen-filled incandescent light bulbs, which General Electric and the rest of the world have been using ever since. General Electric's profits soared, and Irving Langmuir became an international celebrity.

Years later he wrote, "Perhaps my most deeply rooted hobby is to understand the mechanism of simple and familiar phenomena." General Electric had given him the opportunity to do just that, and the fact that the simple phenomenon was a practical one—unwanted black deposits on light bulb housings—did not detract from his satisfaction. The next phenomenon he chose to study, the oil-on-water research to be described in this chapter, probably did not benefit General Electric directly but did lead to a Nobel Prize for Langmuir.

Scientifically, all of it has to do with monolayers: first, in relation to light bulbs, layers of single atoms on tungsten filaments and, once that work was done, an easy transition to layers of single molecules on a water surface—essentially the same experiment that Ben Franklin and Lord Rayleigh and Agnes Pockels had done. Langmuir's *America* would have been unimaginable to Ben Franklin. Even the uses to which electricity had been put would have taxed his credulity, for Franklin, when he was an "electrician," dealt only with static electricity; the ability to make electric charges flow as "currents" was not developed until much later. But Langmuir's *surface chemistry* would have given Franklin no problems at all. Langmuir's experiments on this topic were Franklin's spreading experiment all over again; measuring the surface areas occupied by known amounts of oil or other surface-concentrated molecules.

Langmuir of course used the indispensable equation already given on p. 78, and a modified form of Agnes Pockels's trough (Fig. 16).

3. Questions of Priority

The essence of Langmuir's work on molecular monolayers is contained in a single paper, published in 1917 in the *Journal of the American Chemical Society*. Aside from its unquestioned brilliance, the paper is a fascinating document because of Langmuir's intense concern to establish his priority in the development of the central idea of molecular orientation in monolayers. W. D. Harkins at the University of Chicago had somewhat similar ideas at about the same time, but Langmuir insisted that he was the first, and his paper begins with a *two-page footnote* devoted entirely to his priority, a kind of explosion of incredulity at the very idea that someone else might actually put forward a rival claim. The words flowed so fast that Harkins's name was distorted to "Harkness" at one place in the paper, and the distortion remained there through readings by editor and proofreaders. Here are some direct quotations:

> The fundamental idea of the orientation of group molecules in the surface and in the interior of liquids as a factor of vital importance in surface tension and related phenomena, occurred to me in nearly its present form in June and July, 1916. . . . Work on oil films began in June. As a result of this extensive work, clear ideas were obtained as to the orientation of molecules in the surface of liquids . . .
>
> An account of this work was given in some detail in a paper read at the New York meeting of the American Chemical Society in September, 1916, and a short abstract was published in *Met.Chem.Eng.*, 15, 468 (1916). . . . On Oct. 27th, at the Cleveland meeting of the American Physical Society, I described the mechanism of surface-tension phenomena, and presented the data for γ_0 for about 100 typical substances . . .
>
> Since this time I have found that Prof. W. D. Harkins has been developing an essentially similar theory of surface-tension phenomena. He has recently published two papers on this subject. . . . Nearly all the data in the second of these papers (except of course the new experimental data presented), had been worked over by me during the summer of 1916 and it was my intention to publish them together with the material now given in the present paper. Dr. Harkness [*sic*], however, kindly sent me advance manuscript of his papers, and I have therefore been able to avoid duplication of his work. Harkins had expected that his two articles would appear simultaneously, and as an unfortunate result he failed

> in his first article to mention my prior publication. In his second paper he refers to my work, but, by an oversight, fails to refer to the publication of my general results in *Met.Chem. Eng.* . . .

(One can almost "see" the anger in Langmuir's face!)

> Under these circumstances it is desirable to quote a part of the abstract in *Met.Chem.Eng.* . . . as follows. . . .

There does indeed follow close to one thousand words of the previously published abstract. And this is not in a private General Electric Company publication, but in America's most prestigious chemical periodical!

It is difficult to imagine that any journal today would allow even a short discussion of priority that has the directness of Langmuir's; certainly no discussion as detailed as his would be permitted. Moreover, today's journal editors would not permit the reprinting of lengthy portions of previously published material for any purpose. It is even difficult to understand how the reprinting could have been permitted in 1917, for we are by then in the modern era, with most of the present rules for scientific publication already in force.

4. Award of the Franklin Medal

The irony of all this is that Irving Langmuir himself was guilty of a failure to give credit where credit was due, the injured party in this case being none other than Benjamin Franklin!

In his 1917 paper Langmuir carefully explains that a small quantity of oil, placed upon a clean surface of water, will spread out until a definite large area has been covered and then will spread no farther. He writes that an article by the French surface chemist Marcelin first drew his attention to "this remarkable phenomenon," and appropriately cites the important contributions to the subject that had been made by Miss Pockels and Lord Rayleigh. But there is no mention of Franklin, no mention of the fact that Franklin had been the first to call attention to how remarkable the phenomenon indeed was. Lord Rayleigh had not been guilty of this omission and made explicit reference to Franklin in his 1890 paper, which was a printed version of a lecture at the Royal Institution. Molecular dimensions were not mentioned there, because there had not been sufficient time at the lecture, and Rayleigh had no need to repeat the historical references in later papers that were explicitly about molecular dimensions. Langmuir quotes only an 1899 paper, and presumably that is the only Rayleigh paper that Langmuir had read.

The truly bizarre part of the story comes seventeen years later. (Had no one in the interim showed at least Rayleigh's reference to Langmuir?) Irving Langmuir, after winning the Nobel Prize in 1932, was now, on May 16, 1934, to be awarded the Franklin Medal by the Franklin Institute in Philadelphia, an organization endowed in honor of Benjamin Franklin, to pay lasting tribute to him as a scientist. It was the first time the medal award ceremony took place in the Institute's recently completed magnificent building. That very morning an informal ceremony had taken place to open the Benjamin Franklin Memorial Chamber to the public.

Dr. H. J. Creighton presided and introduced Langmuir with customary eulogy for "numerous fundamental researches in chemistry and physics, especially his work in the domain of surface chemistry." But even Creighton did not seem to be aware of the singular appropriateness of this research in relation to Benjamin Franklin. Langmuir's presentation lecture, published later that year in the Institute journal, was entitled "Mechanical Properties of Monomolecular Films." As would be expected, much of the paper is devoted to films of oil on water. In one place he says, specifically with reference to such films,

> When concentrations from 0.001 to 0.006 are used, the films become so thin that iridescent colors appear.

Remember Franklin's similar words?

> But when put on water it spreads instantly many feet round, becoming so thin as to produce the prismatic colors . . .

Yet there is no reference to Franklin in Langmuir's paper.

What about the audience? All the Institute's officers must have been there, and many of them surely were steeped in Franklin lore. In 1931 the University of Pennsylvania Press had published a book, *The Ingenious Dr. Franklin*, by Nathan Goodman, which reprints a large part of Franklin's 1774 paper in a chapter called "Effect of Oil on Water." Presumably much of the audience had read the book. Did no one at the conclusion of the lecture rise to his feet and say, "But Dr. Langmuir, aren't you forgetting something?" If so, there is no record of it.

5. Molecules at the Interface

Scientifically, Irving Langmuir represents the pinnacle of surface chemistry. He was equally adept at experimental work and theoretical comprehension. He had a rigorously disciplined mind. And he had something special that really distinguished him from all his predecessors, a superb imagination. Applied to molecular surface areas, it enabled him to "see" atoms and molecules with a detail heretofore unimagined. Others assumed that molecules were spheres or undefined "blobs." Langmuir could imagine their overall shapes, and what part was in the water and what part stuck out into the air, and where the molecules were kinked and when they straightened out.

Figure 20 provides the chemical structural formulas of the simplest of the molecules we are discussing here, with a hydrophilic group at one end, attached to a hydrophobic domain in the form of a single long hydrocarbon chain, containing up to twenty-six carbon atoms. It might be thought that these molecules present no need for

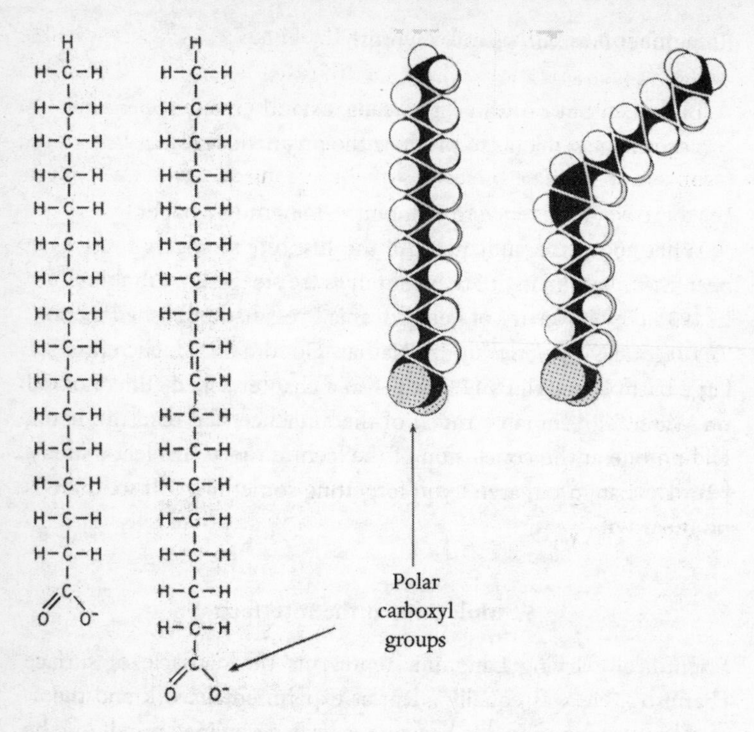

Figure 20. Chemical formulas for the ionized forms of palmitic acid (sixteen carbon atoms) and oleic acid (eighteen carbon atoms). The photographs represent space-filling models for *maximally extended* palmitic and oleic acid molecules (in ionized form). The shaded balls represent the hydrophilic oxygen atoms at the ends of the chains, and the white balls represent hydrogen atoms.

The figure shows that the chemical formula for a hydrocarbon chain, intended to show clearly which atoms are linked to each other, exaggerates the length of the chain. The carbon atoms in an actual chain are staggered along the chain direction, as shown by the molecular models.

The oleic acid molecule contains a "double bond," which means that it has two hydrogen atoms fewer and more electrons shared between the two C atoms so bonded. The "double bond" prevents internal rotation and confers a mandatory kink on the chain at that point. Compare figure 11.

imagination, and that the pictures in the figure provide a reasonable basis for guessing molecular shape. But that is not true. The structural formulas are only representations of how atoms are linked to each other and cannot tell us anything about the disposition of the atoms in space. The molecules might be long and thin, imitating in nature the way we write the formulas on paper. But it makes equally good sense to think that the molecules might be coiled up into compact balls to occupy minimal space. (Protein molecules with incredibly long chains—thousands of atoms—in fact do that.) Without experimental measures of molecular shape, who can tell?

We can illustrate Langmuir's solution to this problem in terms of his results for the fatty acids. Fatty acid molecules have long hydrocarbon chains with a strongly polar "carboxyl" group (COOH) at one end, as illustrated by the chemical formulas in figure 20. In contact with water the carboxyl groups tend to become ionized—which means conversion to COO^- and release of H^+ ions into the water. (Release of H^+ is in fact the property that *defines* what an acid is.) This process does not continue to completion because the H^+ concentration in the water will not rise above a fairly low "equilibrium" value, but it is certain that some of the COOH groups in Langmuir's experiments must have been in the ionic COO^- form. Langmuir appears not to have been aware of this—at least he does not mention it—but the neglect of ionization is not important here where all that matters is that the terminal group of the fatty acid molecule should be strongly hydrophilic, a condition that COOH satisfies as well as COO^-.

The value of Avogadro's Number was, of course, by this time established, so that the exact number of molecules in a given mass of any acid was accurately known. This means that the conversion of the measured surface area per gram to the surface area *per molecule* could now be considered free of any error, provided only that all of the added acid was actually at the aqueous surface, something that was easily checked experimentally.

Some of Langmuir's results are shown in table 5. The first

Table 5. Molecular Dimensions from Surface Area Measurements

	Cross-sectional area in units of 10^{-16} cm^2	Square root of same in units of 10^{-8} cm	"Length" in units of 10^{-8} cm
Palmitic acid (C16)[a]	21	4.6	24
Stearic acid (C18)	22	4.7	25
Cerotic acid (C26)	25	5.0	31
Tristearin	66	8.1	25
Oleic acid (C18)[b]	46	6.8	11.2
Triolein	126	11.2	13

[a] Numbers in parentheses are numbers of carbon atoms in the hydrocarbon chain.
[b] With "double bond." See figures 11 and 20.

column gives the name of the acid and the number of carbon atoms in the hydrocarbon chain, and the second column gives the measured area per molecule. From the relation between volume, area, and thickness, we can calculate the thickness of the layer of acid on the surface, which (once we have assumed that we are dealing with a unimolecular film) is equivalent to the "length" of each molecule in a direction perpendicular to the surface, and this figure is given in the last column. One other important number can be calculated by taking the square root of the surface area per molecule, and it is given in column 3. If the molecules had an exactly square cross section, then the square root of the area would correspond to the length of the side of the square; in other words, it would be an exact measure of the cross-sectional "width" of each molecule. However, it is easy to convince oneself that the number in column 3 would have essentially this same meaning for other possible molecular geometries. For a circular cross section, for example, the number would be within a few percent of the cross sectional "diameter."

I have deliberately put "length," "width," and "diameter" in quotes here because we are not really *measuring* these distances.

The calculation is what scientists call a model-dependent calculation, influenced by what our minds imagine to be going on at the site of measurement—in this case our conviction that we have a monolayer of molecules at the water surface, with the carboxyl group sticking into the water and the hydrocarbon chain extending outward from it. It is conceivable that the results obtained from the calculation might have been absurd, which would have required abandoning the model, rethinking of the whole chain of argument that we have presented in earlier chapters. The results in the table are, however, not absurd, but reasonable—which is significant support for the correctness of the underlying model.

Langmuir's results (interpreted according to the monolayer model) clearly require the adsorbed molecules to be far from spherical, in fact very asymmetric, with a vertical extension from the surface that is much larger than the cross-sectional diameter in the plane of the surface. In agreement with prediction, the cross-sectional diameters (close to 5×10^{-8} cm) are about the same for all of the fatty acids, but the length increases with the number of carbon atoms in the chain. (There is an additional measurement for a C30 chain with an OH head group instead of COOH, which has a length of 41×10^{-8} cm.) In tristearin, where we have three stearyl chains attached to a single polar head-piece (Fig. 21), the length is the same as in stearic acid, but now the cross-sectional area is three times as large.

An additional conclusion of a somewhat different kind is obtained from the fact that the molecular lengths do not increase in direct proportion to the number of carbon atoms, and that they are somewhat shorter than one might have guessed on the basis of the known distance between carbon atoms in diamond crystals. Langmuir concludes correctly that hydrocarbon chains should be regarded as "extremely flexible."

But what about oleic acid and triolein, where, as noted in figures 11 and 20, there is a "double bond" in the middle of the hydrocarbon chain? The cross-sectional area for oleic acid is twice as large

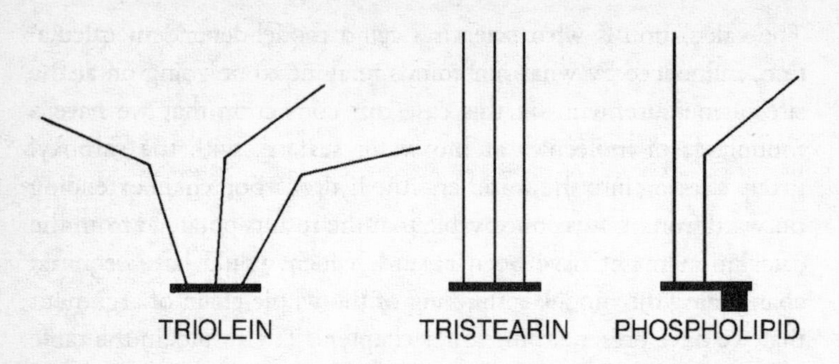

Figure 21. Fats, oils and phospholipids. Schematic representations of triolein (principal component of olive oil), tristearin (typical "saturated" animal fat), and a phospholipid molecule. The straight and kinked lines represent the hydrocarbon chains of oleic acid (shown in Fig. 20) and stearic acid. The latter is like palmitic acid in Fig. 20 but has eighteen C atoms instead of sixteen.

The hydrophilic base is the same for all three molecules, and it has three points of attachment for side chains. Three oleic acid chains are attached in triolein. Three stearic acid chains are attached in tristearin. Only two fatty acid chains are used in phopspholipid molecules, the third position on the hydrophilic base being occupied by another strong hydrophilic group. Many different kinds of fatty acid may be used in phospholipid molecules. Typically one of them has a double bond and the other does not, as in the illustration in this figure.

as for stearic acid, though both molecules have hydrocarbon chains with the same number of carbon atoms. There is a similar difference between triolein and tristearin. Langmuir provides evidence from solubility studies of other substances with "double bonds" that such bonds lead to a small increase in the attraction between water and hydrocarbon, not enough to make the molecule as a whole want to immerse itself in water, but perhaps enough to give the hydrocarbon chain a tendency to lie flat on the surface of the water, instead of extending vertically from it into the air.

He then proceeds to prove his hypothesis by compressing the film, decreasing the surface area by means of the movable barrier B of the trough shown in figure 16. Langmuir's version of the trough contained a device for measuring the lateral pressure in the surface as the area was being reduced and the molecules were being pushed closer together. For palmitic and stearic acids Langmuir found that a pressure increase had only a small effect on the measured areas, consistent with what might be expected if the molecules were initially standing more or less straight up, but (because of chain flexibility) not packed together to their maximal closeness. In the case of oleic acid, however, compression had a large effect, easily decreasing the area by close to a factor of two. The result was consistent with the idea that the hydrocarbon chains were in fact lying on the surface and had to be slightly coaxed to stand straight up. After coaxing, the surface area per molecule was comparable to that of stearic or palmitic acid.

In the discussion of his results, Langmuir beautifully demonstrates his professional virtuosity. There is an unwritten rule of scientific research that should be invoked whenever a new theory is proposed to account for some experimental phenomenon. The rule is that if there is an *obvious* experiment that should be done next, something that the theory predicts that should be confirmed, then this experiment should be done at once, preferably before the theory is published.*

* The rule is primarily directed against fanciful ad hoc explanations. Regrettably, the rule is almost forgotten today in many sectors of the scientific community. Innovative ideas are welcomed purely for the sake of their novelty. The innovator dwells briefly in the limelight, admired for his ingenuity. It is considered uncharitable to demand that an "obvious" test be carried out. Meanwhile, a better scientist in the same area of research can become frustrated because he is put in the position of having to defend his own work against unsupported flights of fancy. This breakdown of traditional scientific discipline, if it becomes the general rule, constitutes a real danger to the future of science. It may tend to drive the best people out of the profession, while the more pedestrian stay on, fruitlessly following false leads.

Irving Langmuir invokes this unwritten rule. His new idea, the logical conclusion from his experiments, is the idea of molecular asymmetry and orientation at the water surface. The evidence for it, however, comes exclusively from experiments with molecules that have very long hydrocarbon chains. There is nothing in the theoretical idea itself to suggest a limitation to long chains—molecules with a similar structure, apart from having much shorter hydrocarbon chains, should also be hydrophobic at one end and hydrophilic at the other end, and should be similarly oriented at air/water interfaces. The "obvious" next experiment is to see whether this prediction holds true. Even a molecule as small as the ethyl alcohol molecule (Fig. 11) should be oriented at the interface with its OH group in the water and its CH_2CH_3 segment sticking out into the air.

To make the required measurements directly is not easy. Alcohol and fatty acid molecules with short hydrocarbon chains have appreciable solubility in bulk water—ethyl alcohol, for example, is miscible with water in all proportions. When the hydrocarbon chains are very long (as for the molecules in table 4) they are practically insoluble in the bulk-aqueous medium, which means that the total amount of the organic substance in the surface film is a known quantity, being simply the total amount placed on the surface. For more soluble substances we cannot know a priori how much of what was added is in the surface. Nor can we use variation in surface pressure as a tool, because, where appreciable solubility exists, pressure is expected to affect the way the molecules distribute themselves between the surface and the bulk liquid.

Langmuir turned to the work of his great American predecessor, Josiah Willard Gibbs, to solve this problem. Gibbs, who had died in 1903, was, apart from the attribute of "greatness," the exact opposite of Irving Langmuir. No strenuous life for him, except in the mental processes that went on in solitude in his study. No vivid pictures in his science, but instead equally brilliant *abstractions*, expressions of universal physical principles (mostly in the form of mathematical equations) that are above pictures, applicable to all matter independently of how we imagine molecular shape and size. He had fame among the cognoscenti, a handful of theoretical physicists around the world, but was almost unknown to the American public. He had none of the national acclaim that Langmuir received.

One of Gibbs's abstractions was a theoretical adsorption equation, from which one could deduce the amount of material in a surface on the basis of how the *surface tension* varies as a function of the concentration of the same substance in the bulk solution. One could then determine how the *amount in the surface layer* itself varies with concentration, approaching a saturation limit, and from this one could calculate the mass and area of a "filled layer." With the aid of these theoretical relations, the measurements of surface area per molecule could be made indirectly by using surface tension measurements. The requisite measurements, it turned out, were actually already in the literature, published in 1891 by the German physical chemist Isidor Traube, so that Langmuir needed to do no new experiments himself. Traube had already demonstrated (but had not explained) a systematic effect of lengthening of the hydrocarbon chain in experiments that involved a series of otherwise similar molecules, such as the alcohols.

Langmuir's application of the Gibbs equations to these data yielded spectacular results. The results demonstrated that molecules with very short hydrocarbon chains obey the same rules as their long-chain insoluble counterparts. The "filled layers" at the high concentration limit are one molecule thick, the cross-sectional area is between 20 and 30×10^{-16} cm^2 per molecule, the molecular "length" is appropriately short, strictly dependent on the length of the hydrocarbon chain.

There is, at least for me, a kind of magic in these experiments of nearly seventy years ago. Molecules are so small, in a way it's a triumph of the human imagination to realize that they must exist at all. But, here, with the simplest conceivable kind of apparatus (in principle, one could have continued to use the pond at Clapham Common), we have so much more, actual measurements of layers that are only one molecule thick, *realistically* interpretable in terms of what single molecules are doing. Today there are more sophisticated tools (especially X ray crystallography) to provide more direct molecular pictures. Molecular models, with atomic positions precisely defined, have become a familiar sight, almost taken for granted, without a thought as to how they were obtained—which, of course, is progress, but, at least to my way of thinking, less exciting than when the human brain had to figure it out by itself, without the aid of advanced technology.

CHAPTER SIXTEEN

Biology—Cells and Membranes

I can find no intelligible ground for refusing to say that the properties of protoplasm result from the nature and disposition of its molecules.

Thomas Henry Huxley, from a lay sermon delivered in Edinburgh in 1868

1. Missed Connections

LIFE began in water, about three billion years ago—the first "thing" that could hold itself together and derive energy from food or sunlight and make clones of itself. And water has remained indispensable to all forms of life. There is life without oxygen—some organisms prefer it that way—but there is never life without water.

It goes without saying that living matter, like any other matter, is composed of molecules. The most prominent among them are special molecules, found in every known form of living matter and thus presumably essential to the living state. It is noteworthy that the molecules that Irving Langmuir used for his research—fatty acids, oils, and fats—belong to one of these essential categories, called "lipoids" early in this century but going by the name of "lipids" today. Their chemical formulas were known before Langmuir's time, and Langmuir used lipids purely as chemical exemplars, as devices to illustrate molecular behavior, without attaching particular significance to their origin in living matter. But biologists knew these molecules, too, for their ubiquitous presence in living matter.

Many years earlier, not long after the publication of Darwin's *Origin of Species*, Thomas Huxley had laid down the ground rules for future progress in biological science: look to the molecules, he had said, their chemistry and their disposition. In 1917, after Langmuir, "disposition" might have loomed especially important, directing attention to molecular orientation at boundaries where lipid and water meet, an orientation dicated by the opposing characters of the hydrophilic and hydrophobic ends of the lipid molecules. One would think that biologists would have been gripped by excitement. Isn't creation of order out of chaos one of the prime attributes of living matter? And here we have molecules that exist in every living cell that *order themselves spontaneously*! A potential bridge between structural chemistry and structural biology had appeared.

But science was then still free in spirit. Like the Israel of old, it had no king: every man did that which was right in his own eyes. The biologists/chemists who might have created this bridge from Langmuir to biology had other things to fascinate them.

What happened in fact was the creation of a separate branch of science called "biochemistry," which chose to focus for many years almost exclusively on a single aspect of the biology/chemistry interface—one that did not involve the possible structural role of lipids. Contacts between the newly arrived "biochemists" and the older "chemists" were at first infrequent, far less common than they are today. At the same time the biochemists had surprisingly little interest in "biology" as a whole—in cells and membranes and muscles and nerves. Thus the part of the scientific community that might have been a bridge became a barrier instead. The connection between Langmuir and biology did not become part of the mainstream of science until very much later.

Geographically it is rather curious. Schenectady, New York, is within one hundred to two hundred miles of what were then America's most prestigious universities. Travel by train (and later by automobile) had become reliable and comfortable. Already at the turn of the century, Willis Whitney, director of research at the General Electric Company, the man who was to hire Irving Langmuir a few years later, commuted regularly between Boston and Schenectady—he initially accepted the position at General Electric only on condition that he could simultaneously retain his professorship at MIT. There is no record to indicate that he found the journey arduous. But ideas, apparently, did not travel as easily as people.

To be fair, it should be pointed out that communication failures occurred in both directions. Irving Langmuir was occasionally invited to lecture at conferences or symposia in England and in the United States, where protein chemists were often present and he thus became aware of the problems they faced in trying to measure and account for the detailed molecular structure of proteins. He became intrigued by the speculative and untenable ideas held on this subject by the now notorious Dorothy Wrinch and he actively supported her in his own papers. Even minimal willingness to listen to some of his colleagues at Harvard or Yale (instead of just lecturing to them) would have quickly convinced Langmuir that Wrinch's models were ludicrous.

This chapter will present a capsule history of the development of knowledge and ideas about the living cell and its contents and membrane, taking us to about 1950. There were of course a few lonely prophets during this period, who (contrary to the prevailing trend) made *proper connections* and sought to emphasize the importance of lipid molecules and their molecular orientation at interfaces. But they were unheeded. Their work was hardly even debated, literally remained unknown to most of the scientific community until much later. I shall devote the following chapters to two of these prophets. Postponing discussion of them in this way gives a better feeling for the true pace of progress in twentieth-century biology—the trials and tribulations that came before the dawn of present understanding—than if they were inserted here in their proper chronological place.

2. Cells Are the Building Blocks of Living Matter

Robert Hooke in 1665 first saw the subdivision of living matter into cells when he looked at cork in his compound microscope, but no one dreamed then that Hooke's pictures revealed a fundamental property of all living matter. Other microscopists about the same time discovered "animalcules," little animals, but again had no conception of the relation between them and larger organisms. In fact, some of the ideas that were then expressed (laughable to us today) were the very antithesis of the notion that living tissue is built up from microscopic "cells." Even quite prominent scientists, looking through the microscope at sperm cells, mistakenly imagined that they could see preformed human figures in miniature and published illustrations of the "homunculus" crouched there, complete with head and legs and arms (Fig. 22).

As men's minds (and the resolution of their microscopes) became sharper, the conception that many living tissues are aggregates of single vesicular bodies, "cells," steadily gained acceptance. The German *Naturphilosoph* Lorenz Oken (contemporary of poet-philosopher Goethe) was quite explicit on the subject as early

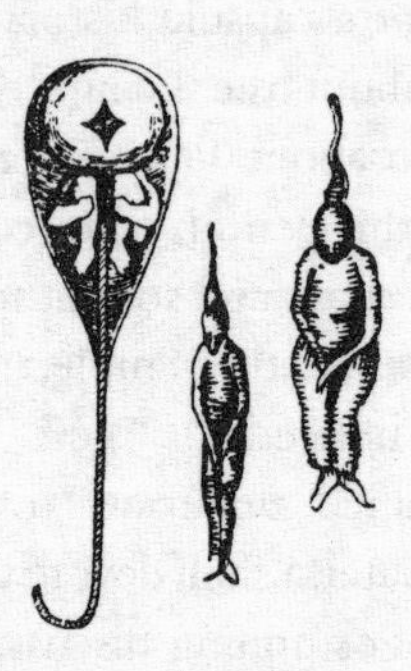

Figure 22. Some microscopists of the 1690s thought they could see the human form in miniature when they viewed human sperm through the microscope. From Charles Singer's *Short History of Biology*, Oxford University Press, 1931.

as 1805: "All organic beings," he wrote, "originate from and consist of vesicles or cells."

But it is Theodor Schwann (1810–1882) who is generally credited with establishing the cell theory as accepted biological doctrine. In 1839 he published his famous "Microscopical researches on the similarity in structure and growth of animals and plants," based in part on his own careful observation of segmentation (i.e., cell division) in the early stages of development of an egg into an embryo. All living matter is made of cells, he stated, or of substance thrown off by cells. The cells have a life that is to some extent their own but subject to regulation by the organism as a whole. All cells share characteristic features: they have a nucleus, protoplasm, and cell membrane. "The great barrier between the animal and vegetable kingdoms, viz. diversity of ultimate structure, thus vanishes. Cells, cell membrane, cell contents, nuclei, in the former are analogous to the parts with similar names in plants." These tenets have remained unaltered to the present day.

Theodor Schwann has other credits to his name, all gained while still a very young man. He coined the term "metabolism," meaning "subject to change," for the chemical reactions that go on inside a cell for synthesis and energy production. He did pioneering work on muscle contraction. He observed that fermenting sugars were teeming with micro-organisms (yeast) and correctly inferred that the micro-organisms were the *cause* of fermentation, anticipating Louis Pasteur by twenty years. Schwann was an altogether brilliant man. Surely destined, one would think, for fame and fortune.

Alas, no! Schwann's career as a scientist did not continue beyond age thirty. The contention that alcohol fermentation is caused by a living organism (made not only by Schwann, but also by a Frenchman, Cagniard de la Tour) ran afoul of the mainstream of contemporary chemical opinion, which championed the cause of "pure" organic chemistry against "vitalism." Germany's most powerful chemist, Justus Liebig, in particular, sneered at the possible intervention of a living creature in fermentation, and his associate Friedrich Wöhler published a vicious caricature in the famous *Annalen*, which Liebig edited. This satire—a report of an imaginary microscopic investigation—described "yeast" as a tiny animal with exactly the same shape as a distilling flask that one would use for the laboratory production of alcohol. It had a funnel-like mouth; the body of the "flask" was imagined as a "stomach"; there was a little tube to represent the anus (for "excretion" of alcohol) and another tube for the escape of carbon dioxide. Elsewhere in the same journal were more serious papers on purely chemical fermentation, pure test-tube production of alcohol, as carried out in Liebig's laboratory.

It is generally believed that Liebig's ridicule caused a timid appointments committee to deny Schwann a professorship at the University of Bonn that he had otherwise every reason to expect. Schwann was sufficiently discouraged (or disgusted) to decide then and there to give up science altogether. He devoted the rest of his life to mysticism and theology, with professorships in theology at Louvain from 1839 till 1848, and at Liège from 1848 to 1879.

3. Membranes Define a Cell

It is hardly possible to entertain the theory that living matter is subdivided into cells (with some capability of living independently of each other) without at the same time imagining that each individual cell must be surrounded by an enclosure or membrane to keep the inside of the cell separate from the external medium around it. But if one tries to define precisely what properties the membrane should ideally possess, then one begins to run into trouble.

One part of the problem is easy. Some of the intracellular components are "colloids," large particles (giant molecules like proteins or DNA) that can be restrained quite simply by means of a "sieve." Almost any kind of membrane can hold these molecules in place, provided only that the membrane doesn't have huge holes in it.

Even some fibrous substance that could form a reasonably tightly woven meshwork at the cell surface would serve the purpose.

Retention of large molecules is, however, not enough. Ordinary small molecules must also be prevented from escaping. William Bayliss, author of an articulate and delightful textbook, *Principles of General Physiology* (1917), put it this way:

> An amoeba, after having taken in a vegetable cell, proceeds to digest the substances contained therein. The products, in order to serve as food, must diffuse from the digestive vacuole into the other parts of the protoplasm. But if they were able to diffuse out from this protoplasm into the water around, they would be lost to the organism. There is good reason to believe, therefore, that there must be some layer or film on the outer surface of an amoeba through which dissolved non-colloidal substances, such as sugar and amino acids, cannot pass.

But "not passing" does not really solve the problem adequately. Bacteria and many other unicellular organisms do not get their food by ingesting other cells, but by direct "infusion" of the small nutritive molecules. How can a membrane that will not let these substances escape nevertheless allow them to enter the cell? This ticklish question made membranes a subject of contention for many years.

The very existence of membranes was questioned. There were some who argued enthusiastically that the protoplasm within a cell—presumably chiefly the macromolecules—might be continuous, self-adhesive, and that the protoplasm boundary could suffice to define the cell boundary. According to this theory, the confinement of small nutrient molecules to the cell interior was explained by tight "binding" of these molecules to the protoplasm.

This idea is actually made untenable by the so-called *osmotic* properties of cells, a term that describes the flow of water into the interior of cells when the salt balance between inside and outside is not right. The water flow leads to swelling and an increase in internal pressure and, if carried far enough, to rupture of the cell, which is something that couldn't happen if there were no cell membrane to rupture. But this theoretical objection was not enough to deter the faithful believers in the protoplasm binding theory for retention. Strangely enough, disciples of this theory, heretics of conventional wisdom, still lurk in the shadows to this very day.

Most sensible biologists did not of course question the *existence* of an actual physical membrane, even though they could not quite figure out how it worked. But until well into the 1950s they looked in the wrong places for explanatory inspiration—even to the purely inorganic substance copper ferrocyanide, which can be induced to form membranelike sheet structures with permeability to water but not to sugar or to many salts, but which in no other way resembles anything that could conceivably occur in a living organism.

4. Dynamic Biochemistry

Biochemists, as I said earlier, had their attention on another subject, and this was "dynamic biochemistry," one of the great scientific success stories of the twentieth century.

To define "dynamic biochemistry" we go back to the dispute between biologist Schwann (and later Pasteur) and chemists Liebig, Wöhler, and others. Schwann demonstrated (at least to his own satisfaction) that a living micro-organism (yeast) is responsible for alcohol fermentation. The chemists called this "vitalism," and insisted that fermentation must be a purely chemical process.

In a sense, both were right. Yeast is indeed responsible for fermentation, and it is indeed a living thing, a unicellular member of the animal kingdom. But it does its work by means of specific intracellular catalysts first named "ferments," but now called "enzymes." These catalysts are themselves molecules, carrying out their catalytic function within the framework of the laws of ordinary chemistry. They can be removed from the yeast and placed into a laboratory container, and there can catalyze the same reactions that they catalyze within the cell. The living organism itself is in fact not necessary, except as a factory for making enzymes, and the chemistry in the test tube is the same as it was in the cell. It is not, however, the kind of chemistry that the *Annalen* satire had imagined, with distilling flasks and retorts. Enzyme molecules are giant molecules, in the class of "proteins," with a complex mode of action that would not have been conceivable in Liebig's day.

The discovery of separable and soluble enzymes led to an unparalleled spurt of activity, lasting until beyond the middle of the present century. Fermentation, the production of alcohol from sugar, was just one of a multitude of examples of biological catalysis. All living organisms, from bacteria to man, were found to employ well-defined chemical pathways by which the working chemicals of life are synthesized and by which they are degraded to yield energy or transmit information. The pathways were found by and large to be similar from species to species, and all species had similar enzymes that could be extracted from the cells without loss of catalytic ability. At the height of activity, dozens of new enzymes were discovered every year.

Even after the discovery of new enzymes slowed down, the chemical structures of the enzyme (protein) molecules had to be investigated, and their working mechanisms had to be understood. Regulatory mechanisms were needed in the cell to put a brake on biosynthesis in order to prevent oversupply, and these mechanisms were found and explained. How are the enzymes themselves synthesized? What goes wrong in disease? The "golden age" of biochemistry, it has been called. It catapulted biochemistry to a position of unparalleled supremacy and influence among the life sciences.

Most of these studies were, of course, made in the laboratory after the contents of biological cells were transferred into test tubes, flasks, or other suitable containers, generally made out of glass or plastic. The *in vitro* chemistry appeared to be no different from the *in vivo* chemistry, which made it reasonable to regard the biological cell itself as simply the living "container," without giving too much thought to how it was constructed and to whether it had properties other than the simple property of containment. The cell membrane? That's the filmy sac in which the chemistry of life is enacted.

Biochemists, entranced by dynamics, gave little or no thought to membrane chemistry at all until late in biochemical history, when it was found that some enzymes were associated with membranes instead of being dispersed in the aqueous cell contents.

5. Chemistry of Lipids

There also existed a group of "lipid chemists," technically "biochemists" because they were working with chemicals from living organisms, but in fact outside the mainstream of biochemistry. There were only a few of them. Michel Eugène Chevreul, an extraordinary Frenchman who lived to the ripe age of 103, published the first textbook on the subject in 1823, and some of the best lipid chemists continued to come from France till the early part of this century.

The work in France established that lipids of living tissues could be divided into two categories, the so-called *élément constant* and *élément variable*. The latter, fats and oils (including olive oil), represented food reserves, stored for future use in specific places in the body and not found elsewhere. They were mobilized for metabolic use under conditions of fasting and disappeared completely when animals were starved to death. The "constant" category, typified by *phospholipids* (Fig. 21), were found to exist in every known cell in animals or plants. They were present in relatively constant amounts regardless of nutritive state. They were still present postmortem after an animal was starved to death. They were clearly different from food reserves, essential to life in some way. But how? What is their function?

Three theories were put forward, and the influence of the purely dynamic view of biochemistry was so strong that two of the theories visualized the essential lipids as participants in the dynamic life of the cell. One theory held that the phospholipids were necessary intermediates for fat mobilization. Its proponents believed that the storage fats could not be used by a cell without intervention from phospholipids, and that the phospholipids remained in place when all the fats were used up. The second theory held that phospholipids might be carriers for oxygen transport within cells (utilizing the "double bond" that most of these molecules possess), a proposal that was speculative to the point of fantasy but that gained adherents because it was "innovative."

Only the third theory saw phospholipids as playing a structural role, and even then there was division as to whether the structural influence extended throughout the cell protoplasm or whether it was confined to the cell membrane. J. B. Leathes and H. S. Raper, in a 1925 book inappropriately titled *The Fats*, present lucid arguments to suggest that the phospholipids might be essential structural components of cell membranes, but their view did not at the time carry more authority than other proposals.

These incompatible ideas were allowed to exist side by side for many years. There appeared to be little incentive to resolve the conflicts between them, perhaps because the majority of biochemists lacked interest in the subject.

6. Water Is Surely an Important Bio-molecule?

A striking feature of Benjamin Franklin's attempt to explain the spreading phenomenon is the natural way he talks about particles of oil and particles of water, and it is quite clear from the context that he means the same thing that "molecules" of oil and "molecules" of water would mean today. Franklin was self-taught, scientifically unsophisticated, without previous experience in thinking about the liquid state. Here, in preparation for an explanation, he began spontaneously by saying to himself, we have oil molecules, poured onto water molecules, and I must think about both of them if I want to understand what happens when they meet. It is perfectly logical, it's common sense—how else could the problem be approached?

One would imagine that biologists would have approached the molecular interpretation of life processes in the same way, just as spontaneously and naturally as Franklin did. Everyone knows that life as we know it originated in the ocean and continues to exist exclusively in an aqueous environment. Nucleic acids, the carriers of genetic information; the multitude of organic molecules that participate in metabolism within the cell; the protein enzymes that catalyze the metabolic reactions—they are all surrounded by a huge excess of water molecules as they carry out their chemical tasks. The extracellular environment likewise is an aqueous solution. Even when cells are pushed tightly together, as they are in certain tissues, they are still bathed in water at more than half of the cell surface. The cell membrane, if one wants to think about it (as I have already said, not many people did), must be viewed as a thin septum between two mighty seas of water.

Water, water, everywhere. It surely can't be ignored when we try to understand proteins and DNA and how they work, or how enzymes manage their diverse catalytic functions? Surely water molecules must be replaced when small "substrate" molecules bind to an enzyme molecule? Is there a tendency for hydrophobic domains of different large molecules to come together to minimize the area of contact with water, and is this a possible *mechanism* for guidance in the complex space of a cell? Surely water must be kept in mind when we try to understand how membranes are formed from their constituent molecular building blocks?

The reality is that water molecules were ignored, not only by biochemists but even by most "pure" chemists who, in the middle of the twentieth century, increasingly turned their attention to problems related to biology. It was not done deliberately. No one pointed to all the surrounding water molecules and gave reasons for why they could safely be neglected. Water just didn't enter the consciousness at all. Just as the cell membrane was a "sac" in which the drama of dynamic biochemistry was enacted, so was water thought of as no more than an inert "solvent" for the true molecular actors in that drama.

The single event that probably did most to turn the tide was the publication in the 1950s of two papers by Walter Kauzmann. He was a physical chemist doing research on proteins and was not interested in lipids or cell membranes. But he came to realize the importance of the aqueous environment for protein molecules, both on the three-dimensional *structures* adopted by these giant molecules—they could have existed as long and extended chains, but instead they tend to be folded into tight compact balls—and on their *function* as biochemical catalysts. In both cases, he pointed out, hydrophobic and hydrophilic domains in the protein's molecular architecture must play a dominant role. Communications were better than they had been earlier in the century and Kauzmann's basic message spread rapidly from his own relatively narrow field of interest to other areas of biological science, such as drug action, lipids, and membranes.

7. Crusader for Water

Quite early in the century there had actually been an exception to the general exclusion of water from people's ideas about biological mechanisms, but the exception is a puzzle in itself, almost as much as the neglect of water by the majority.

The exception was Lawrence J. Henderson, a Harvard professor with red hair and red beard, pink cheeks, and bright blue eyes. He was a maverick, a mystic, a man with little patience for rigorous rules of scholarly evidence, essentially the antithesis of the typical academician. He was a man with charisma, but not much in the way of scholarly accomplishment. In a book that won him international acclaim, *The Fitness of the Environment*, he did indeed extol the virtues of water in biology with almost religious fervor, but not, it turned out, with much success. It was a curious kind of exposition, a praise of water *in principle*, a sort of philosophical exercise, but without substance.

Henderson held an M.D., but never practiced medicine. He taught courses in biochemistry and in physiology and is said to have been an inspiring teacher, emphasizing the broad perspectives: his students and coworkers remember him with admiration and affection. As early as 1911 he started to teach a course in the history of science, perhaps the first such course in America. By 1926 (when he was still not yet fifty) he perceived a similarity between human interactions in society and molecular interactions in blood and began to teach a course in sociology based on that idea. Concurrently, throughout his career, he was active in university affairs, a powerful influence in charting the course of Harvard's academic future.

As befits a man of broad perspectives, Henderson was convinced of the need to bring the knowledge of physical science to bear on the problems of biological or medical science. His influence on Harvard's administrators made it possible for him to put this conviction into practice, and thereby he indirectly affected my own career. He had in 1920 established a Department of Physical Chemistry at Harvard Medical School, an unprecedented venture for a medical college. In 1947, with a fresh Ph.D. in physical chemistry from Princeton, I wanted to make a switch to biological science, and the Harvard laboratory was the logical place for me to go. Henderson was no longer alive, but the Department of Physical Chemistry was very much alive, under the direction of Edwin J. Cohn, a man of superb organizational skill, who had made the department a Mecca for protein chemists throughout the world. Edwin Cohn was able to provide a position for me, and I plunged forthwith into the turbulent waters (no pun intended) between biology and physical chemistry, where I have been ever since. Without Henderson, there would have been no laboratory of this kind for me to go to—even with the Harvard laboratory as example other universities did not follow Harvard into so explicit a cross-disciplinary venture.

In contrast to his vision in administrative affairs, Henderson was a strange figure when viewed as a pure scientist. He made some useful contributions to blood chemistry in the 1920s and stimulated others who subsequently became leaders in blood chemistry research. But more often he seemed to be more preacher than scholar; *The Fitness* of the Environment reads more like a sermon than a scientific text. The book was in fact a "revelation," which entered Henderson's mind one day as he was walking down the slopes of Mount Monadnock in nearby New Hampshire. He assembled the facts to support his thesis in a few days (if he knew of contrary evidence, it was not mentioned), and at once set about the task of writing. The entire project required no more than two months from start to finish, and that included interruptions for course lectures and a family funeral.

The word *fitness*, in Henderson's book, is used in the Darwinian sense, that which favors natural selection, and the principal message seems to have been that adaptation is a reciprocal process, the environment adapting to life as well as the other way around. In Henderson's own words,

> How, for example, could man adapt his civilization to water power if no water power existed within his reach?

A circular argument? I am no expert and hesitate to express an opinion, but I cannot help but be reminded of Tweedledum and Tweedledee:

> If it was so, it might be; and if it were so, it would be; but as it isn't, it ain't. That's logic.

And, perhaps, that's right.

The book itself is fascinating reading. It begins with reference to a remarkable series of treatises financed in the 1830s from a bequest by the 8th Earl of Bridgewater. The overall series title was *Bridgewater Treatises on the Power, Wisdom, and Goodness of God as Manifested in the Creation; illustrating such work by all reasonable arguments, as for instance the variety and formation of God's creatures in the animal, vegetable and mineral kingdoms; the effect of digestion and thereby of conversion* . . ., and so forth in the same vein. Henderson was particularly enthusiastic about the seventh volume in this series, *Astronomy and General Physics Considered with Reference to Natural Theology*, written by a famous Cambridge educator, William Whewell, which included a hymn of praise to the properties of water.*

Henderson cites Whewell's list of the remarkable properties of water and makes some additions of his own. He cites the familiar fact that the density of ice is less than that of water, which permits the ice of winter to float on top of water and thereby to receive the warm rays of the sun early in the spring. He cites water's high heat of melting and of evaporation, which, among other benefits, minimize the volume of body coolant that we require to remove heat generated by metabolism. He mentions the then fairly new knowledge that salts in an aqueous medium (but not in most other solvents) are largely dissociated to charged ions; in his own words, "ions are evidently a real contribution to the richness of the environment." Evidently? He does not provide any clue as to what he thought their contribution might be.

Henderson's greatest enthusiasm is reserved for the solvent power of water, which almost nothing can resist, as exemplified by

* Whewell was Master of Trinity College from 1841 to 1866. He was a giant in body as well as mind, and is reputed to have jumped up all the steps of the Hall at Trinity College in one leap, a feat seldom accomplished. It appears that he saw several undergraduates unsuccessfully making the attempt, whereupon "he clapped his cap firmly on his head, took the run, and reached the top of the steps at one bound."

testimony from the geological record: "Under the action of water, aided, to be sure, in many cases by dissolved carbonic acid, every species of rock suffers slow destruction." Thus the oceans contain in dissolved form almost every chemical element, blood serum can dissolve almost every substance that we need to circulate to the tissues, and urine can dissolve an infinite variety of waste products.

But if water dissolves everything, where would life be? Henderson, though nominally at least a part-time physiologist, does not mention cells and membranes, seems not to have thought about the need for sturdy cell walls that water *must not destroy* by dissolution. Given his penchant for crossing disciplinary boundaries, Henderson might have been just the man to have made the connection between Langmuir's work and cell biology—the connection that most biologists did not make until thirty or forty years later—but he missed the opportunity. No novel vistas appear in Henderson's work. He looked backward to old man Whewell when he might have looked forward; he lagged behind less famous and unappreciated colleagues who in reality understood the biological world much better than he did.

CHAPTER SEVENTEEN

Ernest Overton—Gentle Genius.

1. Prescient Intuition

We go backward in time here, but forward in understanding. The preceding chapter has given mixed reviews to biological science around the turn of the century. There was brilliant progress in some areas, but neglect and confusion in relation to cell membranes. Charles Ernest Overton was an exception, someone whose comprehension of cell membrane composition and function was never confused. On the contrary, "prescient" would in retrospect be a good description. Overton's work is outside the main theme of this book, in the sense that he himself never experimented with films of oil on water surfaces, but it establishes a physiological context and shows that some areas of physiology were crying out for just the kind of insight that surface film measurements could provide.

Overton is worthy of mention for another reason; he is an outstanding example of an unsung hero. His work marks him as a great thinker and experimenter, but he was not recognized as such by his contemporaries, which (since Overton lived till 1933) included people like Langmuir and Henderson. Runar Collander, a plant membrane physiologist, describes him as follows in a short biographical sketch:

> A gentle and placid man, Overton had a striking intuitive ability to recognize the great, fundamental problems and to enviseage a means of solving them without recourse to com-

plicated apparatus. . . . He was one of those scientists whose stature is more obvious after their death than it was during their lifetime.

Overton was born in Cheshire, England, in 1865. His father was a clergyman. His mother, a distant relative of Charles Darwin,

Figure 23. Charles Ernest Overton. This photograph is taken from a short biography by T. Thunberg.

moved to Switzerland for the sake of her health in 1882, and took her childen with her. Ernest entered the University of Zürich in 1884, and received a Ph.D. in botany in 1889. He remained there on the faculty for two more years, and then moved to the University of Würzburg in Germany. In 1907 he was appointed to the chair of pharmacology at the University of Lund in Sweden, and in 1912 he married Dr. Louise Petren, a member of a well-known Swedish academic family. Overton remained in Lund for the rest of his life, though he never became fully fluent in the Swedish language and always remained something of an outsider. His most important scientific work was in fact all done before he went to Sweden and was published in German. There is nothing in the published record to indicate that he ever returned to his native England.

2. Permeability of Cell Membranes

Overton's involvement with membrane permeability happened accidentally. His doctoral thesis and his earliest research after that concerned the mechanism of heredity in plants, but he needed substances that could penetrate rapidly into plant cells for some of the experiments he had designed. He quickly discovered that a considerable variety of neutral (nonionic) molecules were able to enter the cell almost unimpeded. This was at first surprising because the conventional view at the time was that the cell membrane was a "semipermeable" barrier, freely permeable to water, but virtually impermeable to most substances that might be dissolved in water.

The conventional view was clearly untrue, and Overton immersed himself henceforth in a systematic study of permeation, investigating both a wide range of substances and a wide range of different kinds of cells. The latter excursion had two consequences. One was that it led to another surprise—extremely different kinds of cells were strikingly similar in their permeation characteristics, suggesting that some common principle might be involved in the molecular architecture of all cell membranes. The second consequence was that Overton extended his work to include animal cells as well as plant cells, and thereby made it relevant to animal physiology and to medicine. This led Overton to propose a theory of narcosis or anesthesia in animals, which was destined to receive wider publicity than his more basic studies of general membrane permeability.

Overton discovered a systematic relation between permeability and general chemical identity, one that will not come as a surprise to the reader of this book, familiar with the concepts of hydrophobic and hydrophilic molecular domains, but which was quite novel to Overton's audience. The lengthening of hydrocarbon chains in otherwise similar organic molecules increases permeability, whereas the presence of polar or ionic groups decreases it. The big question was how this could be explained. Overton's answer was to point to a parallel between membrane permeability and *solubility* in fatty substances, which he then went on to test by systematic investigation of solubilities in a model fatty liquid, for which he chose the substance already familiar to us, olive oil. A striking correlation was discovered, which led to several more or less obvious conclusions:

1. There must be some resemblance between a cell membrane and olive oil. The membrane must be impregnated with molecules of the same class as the olive oil molecule—either fats or lipids. Overton explicitly suggested cholesterol or cholesteryl esters, in combination with lethicin, which (though simply intended as an intelligent guess) is actually remarkably close to the truth as we now know it.

2. The correlation with a "solubility" measurement suggests (perfectly reasonably) that the way molecules get across the membrane is by first dissolving in the membrane interior, then diffusing across the membrane thickness, and finally entering the aqueous

medium on the other side. And that the ability to dissolve in the lipidlike interior is the critical factor, limiting how fast the overall process can occur.

3. This mechanism provides, for the first time, a simple explanation for the vital function of the membrane—enclosure and confinement of the cell contents. The molecules that need to be confined are mostly water-soluble, strongly hydrophilic. Their solubility in olive oil is slight. By inference their solubility in the membrane interior is also low, and this explains the low rate of escape from the cell.

Overton's hypothesis was published in 1899, but only in preliminary form, the intention being to publish details in a more extensive publication later. The larger work was, however, never finished. The correlation between permeability and solubility was not quite perfect (molecular size is a secondary factor, for example), and Overton's proposal met with doubt and even violent opposition. The most prominent rival theory was a filtration or sieving theory, in which molecular size was espoused as the *primary* factor and the supposedly "oily" nature of the membrane was denied. Overton's temperament made him averse to combat, and he never replied to the attacks against him.

Only the gradual accretion of experimental evidence over many years has eroded the opposition and has vindicated Overton. Scholarly work and controversy on the subject continue to the present day. A new study designed to defend the theory was published by a very modern "biophysical" physiologist (Alan Finkelstein of the Albert Einstein College of Medicine) as late as 1976. In this study the secondary role of molecular size was taken into consideration as a determinant of the relatively minor diffusion factor (from one side of the membrane to the other) after dissolution has occurred. Not unexpectedly, this work prompted new opposition from those who believed that the size factor had been relegated to excessive insignificance.

3. "Uphill" Transport

Overton's realization of the physiological need for special devices in a cell membrane, different in chemical character from the preponderant fat or lipid, is a marvelous illustration of his "intuitive ability to recognize the great, fundamental problems."

The "fundamental" point here is that while it is true that a membrane's selectivity in permeability explains why vital molecules are prevented from leaking out of a cell at a significant rate, that alone cannot be enough to satisfy the total physiological requirement of transport function in membranes. The same or similar molecules (food molecules in particular) need to be allowed to enter the cell when needed, and thus require some sort of *one-way* passage across the membrane. Overton recognized and addressed this problem right from the start of his work. He understood that the permeabilities he was measuring were "passive," related to simple solubility, without directional preference.

The problem is actually deeper than just the question of how the membrane can transport something in one direction and not in the other. Inward transport often needs to be "uphill," from low concentration to high concentration, whereas passive permeability (or any other *spontaneous* movement) can only go "downhill," from high concentration to low. To go in the "uphill" direction requires expenditure of metabolic energy, specially designed engines to use metabolic energy for this unique purpose. Overton fully understood this. "This must be a phenomenon quite different from the simple diffusion of substances through the protoplasts," he said.

In this comprehension of the fundamental energy requirement for initially creating compositional differences (which impermeability subsequently maintains), Overton was far ahead of his times. It would be at least another generation before this subject became part of the mainstream of biological thinking.

4. Molecular Basis for Anesthesia

Another offshoot of the permeability studies was a hypothesis for the molecular mechanism of narcosis or anesthesia, the present-day fine distinction between these terms being unimportant in this context. Many of the molecules that produce insensibility consist of polar chemical groups with long hydrocarbon chains attached to them—for example, the long chain alcohols, in which the polar group is simply an OH group. The narcotic or anesthetic potency of a series of such molecules with the same polar moiety, but with hydrocarbon chains of different lengths, changes in parallel with membrane permeability and solubility in olive oil, suggesting that solubility in lipids (by implication, in cell membranes) is the core process underlying the anesthetic effect. Overton proposed this view and incorporated it in his most influential work, a book published in 1901 with the title *Studien über die Narkose*. A German pharmacologist, Hans Horst Meyer, reached much the same conclusion about the same time as Overton, though it was less well documented by experimental measurements. Since the two proponents had arrived at their conclusions quite independently, the theory was called the "Meyer-Overton" theory and this name is used to the present day. (Meyer, incidentally, had a son who became a famous polymer chemist, and another son who was a prominent surgeon. A grandson is a well-known physicist today.)

The Meyer-Overton theory was strenuously opposed by Isidor Traube, a controversial and contentious German physical chemist who made a point throughout his long career of writing polemical papers in opposition to various well-reasoned (though at the time not fully proven) theories. For example, he opposed Arrhenius' famous theory that common salts exist in aqueous solution as virtually fully dissociated ions.

We have met Traube before in chapter 15. He had in the 1890s discovered that molecules with polar heads and long hydrocarbon tails, when added to water, migrate to the surface of the water and greatly reduce the surface tension; and that their potency in doing so is a regular function of the length of the hydrocarbon chain. Irving Langmuir used Traube's results in addition to his own for his analysis of molecular orientation. It is this work that formed the basis for Traube's opposition to the Meyer-Overton theory—he argued that narcosis must be a surface phenomenon unrelated to the solution of narcotic agents in olive oil or anything else. At the time there was no reason to reject Traube's hypothesis outright, but he produced only rhetoric to argue its merits, and he did that for many years, well into the 1930s.*

It is possible that the existence of this controversy about anesthesia, and the absence of strenuous rebuttal by Overton, may have contributed to the suspicion and neglect of all of Overton's work.

5. The Farish Syndrome

Is there perhaps an analogy to poor impressionable Mr. Farish here? Are scientists susceptible to the same disease as naive laymen, often believing without understanding, not even trying to think for themselves? Perhaps the disease is more prevalent than we realize. Perhaps we should give it a name, the *Farish Syndrome*, the disease of fallacious enthusiasm for an idea or (worse still) for a person, on the basis of a flagrant misreading of the message that some praiseworthy or even brilliant investigation may carry.

*The molecular basis for anesthesia is still an unsolved problem in the 1980s. Traube's theory no longer has adherents, but there are still enthusiasts for the unmodified Meyer-Overton theory. The most reasonable view today (given other things we know) is that the site of action of anesthetic molecules may be specific protein molecules anchored in cell membranes. In that case, dissolution of the anesthetic in the membrane lipid might still be a prerequisite for access to the specific protein sites, which would lead to retention of part of the Meyer-Overton mechanism. No one has so far identified any particular protein as an anesthetic "receptor," which is why the overall mechanism is still very much in doubt.

Farish had heard about Franklin and his friends on Derwent Water (probably in a rowboat), solemnly tipping a teaspoonful of oil onto the water, and watching the oil spread and still the waves. Convinced that it had really happened and that the report was true, Farish responded quite inappropriately—with increased faith in old Pliny, that often discredited voice from the ancient past. Was there similarly inappropriate and unthinking reasoning in the 1920s and 1930s about Overton and Traube?

Farish reasoned as follows: Pliny said oil stills the waves, Franklin verified it, and therefore it must be true; Pliny also said that vinegar stills tempests in the air, *therefore* I now believe that, too. A similar reasoning here might be: Traube claimed certain molecules migrate to the surface and thereby reduce surface tension, Langmuir verified it, and so it must be true; Traube based a theory of anesthesia on that migration, *therefore* I believe that theory, too.

And believing that theory means believing that Overton was wrong in his theory of anesthesia and, by extension, most likely in everything else he said.

CHAPTER EIGHTEEN

Gorter and Grendel: A Factor of Two

1. Progress Postponed

AFTER Langmuir, was there more to be learned from spreading oil on water and measuring surface areas, using the same simple rectangular trough that Pockels and Langmuir had used? As it turned out, there was indeed new and important information to be gained for biology by applying Langmuir's method to molecules taken directly from living cells. This chapter deals with a single experiment in this category, a beautiful and simple experiment, exactly right for clearing the air of foggy conceptions about cell membranes.

The experiment was conceived by a great Dutch pediatrician, Evert Gorter, and was carried out in his laboratory at the University of Leiden. The result of the experiment was in essence unambiguous. One could quibble about minor details, but they could have been quickly and easily cleared up. Controversy should have ended, and the stage should have been set for a burst of productive research on membranes and their properties. But that is not what happened. Gorter's paper was ignored, even to some degree by Gorter himself, and membranology as a whole continued to flounder in uncertainty. Gorter's conclusions eventually became established "fact," but only forty years later, and they did so then through much more complicated experiments than his own. It is only since about 1970 that people have recognized that the "fact" had been there all along for anyone with open eyes to see.

It is appropriate to comment on the slow acceptance of Gorter's result as well as the result itself. This chapter then divides itself naturally into three parts: a brief sketch of the life of this remarkable pediatrician turned basic scientist; an account of the experiment itself and the possible motivation for doing it; and, finally, an inquest on the aftermath. Why in this age of rapid dissemination of information did it take 40 years to appreciate the message?

2. Evert Gorter. A Life of Combat

Evert Gorter was almost exactly Irving Langmuir's contemporary. They were born in the same month, February of 1881, and both lived until the mid 1950s. Like Langmuir, Gorter was a fighter, but his was not a battle of the strong, fought with the backing of a powerful corporation, as it had been for Langmuir. His battle was rather a lonely fight for survival, against difficult odds, with the establishment usually arrayed against him.

Born in Utrecht, Gorter carried out medical studies at the University of Leiden, receiving his doctor's degree in 1907 with a thesis on tuberculosis bacillus. His first battle concerned his medical career. He wanted to be a pediatrician, to specialize in the treatment of children, but pediatrics was not then regarded as a distinct specialty in the Netherlands, and Gorter had to fight the medical establishment to have it recognized as such. Generous friends placed at his disposal a couple of houses in a low-income neighborhood. There he built a little hospital, and only its success persuaded the university to create a now-famous children's clinic, and (in 1923) to appoint Gorter as professor of pediatrics.

Gorter's second battle was to fulfil his desire to have a laboratory where he could carry out basic research at the same time as being a full-time clinician. He was not allowed to do this until he attained professorial rank, and he was forty-two years old by then. Is there any other modern scientist who began his career, without previous experience, so late in life?

Gorter's most courageous battle was his battle against disease.

Figure 24. Evert Gorter. This portrait hangs in the *aula* of the University of Leiden, among portraits of other famous professors, spanning a history of over 400 years.

He suffered severely from rheumatoid arthritis, a familial disease which had also afflicted his father. His father, like Evert, had a medical degree, but on account of the disease he never actually practiced medicine. Evert somehow conquered the pain and the crippling effects of the disease. He eschewed social events and recreation, but he managed to work full time until the very end of his life. In fact, he really worked more than full time, for he pursued two parallel careers, one in the clinic and one in the laboratory. In the words of Gorter's obituary in *Acta Paediatrica*: "If the spirit ever dominated the body, it did so here."

Gorter earned worldwide distinction in his clinical field. He is considered to be one of the pioneers, one of the first to regard a child in the medical sense as something other than a miniature adult. He founded the discipline of pediatrics in Belgium as well as in the Netherlands, and for some years he was simultaneously professor at Ghent and Leiden, commuting between the two universities. He was founder of one journal of pediatrics and editor of another. He was an honorary member of the American Pediatric Society. But he also had an international reputation as a physical chemist and, as we shall see later in this chapter, probably never got the full acclaim in that area that he merited.

As his physical disability became worse and confined him to a wheelchair, his wife (his second wife, formerly his secretary) helped keep him mobile and active. A former coworker has described her as a "camel driver," commandeering toward subordinates, difficult to get along with. This coworker has also described the typical Gorter day. He was driven to the institute every morning by car and was then helped into his wheelchair. In the morning he went on rounds to see his patients in the pediatric wards. After that he disappeared into his office, where he would read for most of the rest of the day. He was a very fast reader, and many of the papers he read gave him ideas for new experiments. When this happened, he would call his assistant into his office, show him the paper and outline the experiment he had in mind. The poor assistant (subsequently a famous scientist in a different field) couldn't keep up with Gorter's pace. Finally he simply set each new suggestion aside, taking no action unless Gorter brought the subject up a second time.

The wheelchair-directed research led to an outpouring of research papers (most of them using the surface balance), the rate being especially remarkable because Gorter was simultaneously publishing many clinical papers for the pediatric journals—there were also trips abroad to lecture and to promote the specialty of pediatrics.

Early on the morning of May 10, 1940, German armies invaded the Netherlands (and Belgium and Luxembourg and France); by May 14 it was all over. Queen Wilhelmina had fled to England, and the government had surrendered. Leiden had been fortunate to escape bombardment, but its normal existence as a medical center and university town had come abruptly to an end.

These events did not interrupt Gorter's flow of personal courage. He fought the occupying Germans from his wheelchair, a constant thorn in their sides, resisting orders and regulations. Had he been a younger man and in good health, he probably would have been sent to a concentration camp. But even the Germans couldn't send such a prominent man (and a cripple besides) to his death. So they banished him instead to a small town on the northern border of Holland, where his resistance could do them little harm.

After the war Gorter returned to teaching, practice, and research; there was no noticeable interruption in the flow of papers. He had intended to read a paper on the subject "heparin and albumin" at a meeting of the Kolloid Gesellschaft in Hamburg in March 1954, but he died shortly before the meeting. The paper had been typed for publication, and a former associate, Dr. W. A. Seeder of Bloemendaal, read it in his place.

3. Source of Inspiration

What motivated Evert Gorter to undertake a second career in basic science is not known. Given his determination to do so, however, and his lack of specialized training, the decision to study molecules at surfaces was a natural choice. Surface science was at the time at its peak of popularity, and for biologists there seems to have been some sort of mystique associated with it. W. M. Bayliss, for example, in the textbook *Principles of General Physiology*, to which reference was made earlier, has a long chapter on "Surface Action" near the beginning of the book. The following quotations from his summary of this chapter give the flavor of it:

> The surface of contact between a liquid and another phase—solid, immiscible liquid, or gas—has properties differing from those of the main body of either phase.
>
> The surface film is the seat of a special kind of energy.
>
> The interface between phases is also nearly always the seat of electrical forces, the origin of which is usually from electrolytic dissociation in one or other of the phases. But the possibility of phenomena akin to those of frictional electricity cannot as yet be definitely excluded.

Bayliss goes on to talk about "adsorption," which in the present context means the same thing as "migration to the surface," increasing the amount of some solute there. If the solute is an ion, then adsorption may create a surface electrical charge, and Bayliss implies that there may be a deeper significance to this than first meets the eye:

> These phenomena of electrical adsorption play a considerable part in the process of dyeing and of histological staining.
>
> Chemical reactions which lower chemical potential are also favored at a surface.
>
> A number of cases are given where adsorption plays a controlling part in phenomena of physiological interest.

(The examples actually cited by Bayliss are trivial, and the word "controlling" here is considerably exaggerated.)

This sort of "talk" about surfaces would inspire a general sense of excitement and curiosity, but would not of itself point to surface area measurements as a potentially productive line of research. What led to Gorter's decision to go in that direction is not recorded. One possible direct precedent may have been the work of Pierre Lecomte du Noüy, whose very name conjures up mystery and romance. He was a flamboyant personality at the then recently founded Rockefeller Institute for Medical Research in New York, where he devoted himself to surface tension measurements, using a device he himself invented (and marketed commercially), called the "du Noüy tensiometer." He seems to have cut an impressive figure for a while. For example, Dr. Alexis Carrel (French-born member of the Rockefeller Institute, winner of the Nobel Prize in 1912, later coinventor with Charles Lindbergh of an artificial heart) said of him that "his explorations have already brought forth important new facts, brilliant hypotheses, and the prospect of future discoveries." Posterity's judgment has been less favorable; one later critic, Henry Bull, characterized du Noüy's results as artifacts and his interpretation of them as "completely in error."

The point is that Lecomte du Noüy's work dealt in part with molecular dimensions of biological molecules—oleic acid and its salts, for example, and the proteins serum albumin and egg albumin—and he published his results in the *Journal of Experimental Medicine*, an American journal that Gorter would most likely have read on a regular basis. (Gorter later published his own work in the same journal.) Lecomte du Noüy also gave copious references to the work of Pockels, Rayleigh, and Langmuir, so that Gorter would have been able to construct his own theoretical background (much as has been done in this book) independently of what Lecomte du Noüy himself may have written. As we shall see, much of the biological community at the time misunderstood the implications of Langmuir's work, whereas Gorter seems to have understood right from the start. Gorter has been described as being fascinated by the scientific potential of surface measurements for biology "as soon as Langmuir published his experiments in 1917."

4. A Truly Classic Paper

Gorter was assigned a research student, by the name of F. Grendel. In 1925 they published two papers, one dealing with surface area measurements on cell membrane lipids, the other describing the first of a long series of studies in Gorter's laboratory that dealt with spread-out protein molecules. The paper on lipids, as already mentioned, was published in the *Journal of Experimental Medicine*.

This paper by Gorter and Grendel is now recognized as a classic, one of the true gems of the biological literature. In a little over three pages of text and one table of results, Gorter and Grendel demonstrate that there is a difference of a factor of two between (a) the area occupied by all the lipid molecules extracted from a red blood cell, measured when spread as a monolayer on water, and (b) the surface area of the membrane of the cell from which the lipid was derived. Gorter and Grendel felt that they could take it for granted that all of the red cell lipid was originally contained in the cell membrane. The conclusion must then be that the cell membrane is a *bilayer*, exactly two lipid molecules thick.

This simple experimental result is supported by a theoretical paragraph. Succinctly and clearly, Gorter and Grendel show that their experimental result is the theoretically expected result, a priori the most probable one. In the following quotation, note that the word "plasma" refers to the aqueous blood fluid in which the red cells are suspended, and note the use of the word "chromocyte" for the red blood cell itself.

> We propose to demonstrate in this paper that the chromocytes of different animals are covered by a layer of lipoids just two molecules thick. If chromocytes are taken from an artery or vein, and are separated from the plasma by several

> washings with saline solution, and after that extracted with pure acetone in large amounts, one obtains a quantity of lipoids that is exactly sufficient to cover the total surface of the chromocytes in a layer that is two molecules thick. Subsequent extractions with ether or benzene yield only small traces of lipoid substances.
>
> We therefore suppose that every chromocyte is surrounded by a layer of lipoids, of which the polar groups are directed to the inside and to the outside, in much the same way as Bragg supposes the molecules to be oriented in a "crystal" of fatty acid, and as the molecules of a soap bubble are according to Perrin. On the boundary of the two phases, one being the watery solution of hemoglobin, and the other the plasma [also, of course, "watery"], such an orientation seems *a priori* the most probable one. Any other explanation . . . seems very difficult to sustain.

The theoretical argument is given in pictorial form in figure 25. The reference to Bragg refers to one of the early uses of X rays to determine structure—in this case the structure of crystals of molecules that have a dual hydrophobic/hydrophilic arrangement similar to that in lipid molecules.

Experimental measurements were made on red blood cells of six different species, to show that the result obtained was independent of the source and size of the cells:

> We have examined the blood of man and of the rabbit, dog, guinea pig, sheep, and goat. There exists a great difference in the size of the red blood cells of these animals, but the total surface of the chromocytes from 0.1 cc [of blood] do not show a similarly great divergence, because animals having very small cells (goat and sheep) have much greater quantities of these cells in their blood than animals with blood cells of larger dimensions (dog and rabbit).

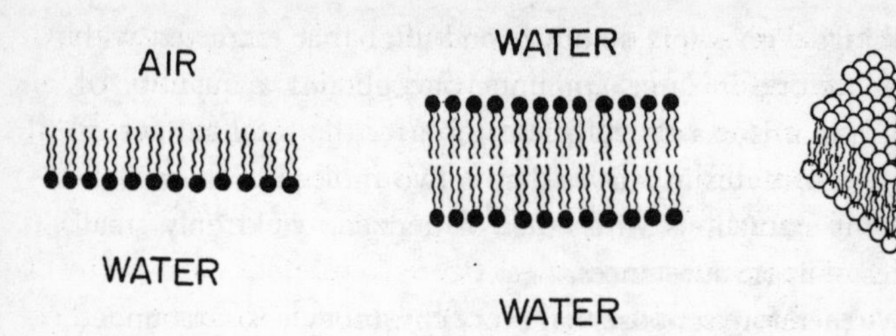

Figure 25. Schematic diagrams of a lipid monolayer and bilayer. Black circles represent hydrophilic ends of the lipid molecules; wavy lines represent the hydrocarbon chains. (The pictures lack a third dimension. Both monolayers and bilayers normally occur as extended surfaces, as visualized in the small membrane patch shown at the right. The hydrophilic ends are here shown as *white* circles.)

The thermodynamic driving force that dictates the orientation of lipid molecules at an interface is the need for the hydrophilic end of each molecule to be dissolved in the water and the simultaneous need for the hydrocarbon chains to be excluded from water. At the interface between two aqueous solutions, hydrophilic groups must face *both sides* of the layer—a bimolecular film is a necessity. The pressure exerted by the water on the two sides holds the two halves of the bilayer tightly together.

The tabulated results, including several different trials for each animal, show that the observed ratios of monolayer area to cell surface area ranged from 1.6 to 2.2. (Some of the results are reproduced in table 6.) The paper ends with a single summary sentence: "It is clear that all our results fit in well with the supposition that the chromocytes are covered by a layer of fatty substances that is two molecules thick."

Why was this result not equally clear to the rest of the scientific community, especially since it is buttressed by sound theoretical considerations and since the proposed ordered arrangement is already known to exist in crystals of fatty acids?

One trivial reason is the brevity of the paper. The work is obviously incomplete. For example, there was no decisive proof that all of the lipid in the experiments came from the cell membrane. If some had come from the bulk protoplasm, then there might have been only enough for a monolayer in the membrane (which is actually most unlikely, because, as figure 25 reminds us, it is one of the results that would be theoretically "difficult to sustain"). Second, the question of the permeability properties of the membrane is ignored. In the case of the red blood cell the rapid exchange of bicarbonate and chloride ions across the cell membrane was by then well known, and that exchange surely required specific transport proteins for these ions within the fabric of the membrane. Gorter and Grendel did not discuss this likelihood, nor how it would affect their calculation.

Table 6. Sampling of Results of Gorter and Grendel

				Surface Area		
	Grams of blood used	Number of cells per mm^3	Single cell in μm^2	All cells in expt (m^2) A	Extracted lipid as monolayer (m^2) B	Ratio *B/A*
Dog A	40	8,000,000	98	31.3	62	2
	10	6,890,000	90	6.2	12.2	2
Sheep 1	10	9,900,000	29.8	2.95	6.2	2.1
	9	9,900,000	29.8	2.65	5.8	2.2
Rabbit A	10	5,900,000	92.5	5.46	9.9	1.8
	10	5,900,000	92.5	5.46	8.8	1.6
	0.5	5,900,000	92.5	0.27	0.54	2
Goat 1	1	16,500,000	20.1	0.33	0.66	2
	1	16,500,000	20.1	0.33	0.69	2.1
	10	19,300,000	17.8	3.34	6.1	1.8
Man	1	4,740,000	99.4	0.47	0.92	2
	1	4,740,000	99.4	0.47	0.89	1.9

But the fact is that few first papers on any subject are complete, and in this case the theoretical justification for a bilayer structure is so compelling that these gaps in the initial experimental support for the model should not have stood in the way of adopting the structure as a viable hypothesis, subject to further investigation and to constant comparison with alternate plausible hypotheses, if such were put forward.

That is not what happened. Gorter and Grendel's result was all but ignored for nearly fifty years. It exerted no influence whatever on the subsequent development of membrane science. Many alternate models were proposed, most of them in retrospect intrinsically untenable, but they were not critically compared to the Gorter-Grendel model.

5. Philistine Doctrine

It would be incorrect to say that any particular model became "accepted" in this fifty-year interim, but one model structure, published in 1935 by James Danielli and Hugh Davson, appears to have received wider dissemination than its competitors and was frequently used to illustrate what a membrane might look like.

The model is shown in figure 26, and one of its attractions was that it includes protein molecules, which are missing from the Gorter-Grendel model. But the protein molecules are shown as coating the outside of the bilayer. How can they in that position fulfil the necessary function of membrane proteins to regulate the transport of metabolites and other hydrophilic species *across the membrane*, from one side to the other? No one apparently asked the question. Furthermore, would not the proteins in their claimed external position interfere with the thermodynamic requirement of the hydrophilic domains of lipid molecules, which we know to be a thirst for water? No one addressed that question either.

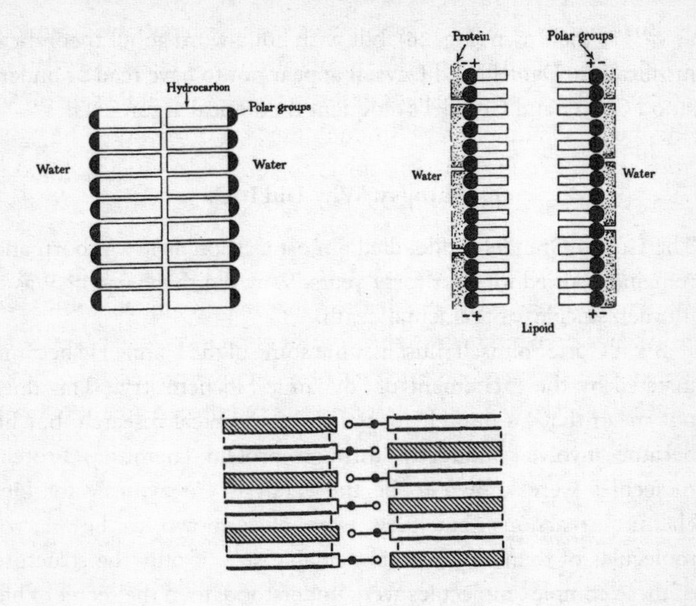

Figure 26. Rival ideas about membrane organization. A and B are two illustrations from a book written by J. Danielli and H. Davson. The titles (quoted verbatim) are A. Structure of red cell membrane (after Gorter and Grendel), B. Structure of cell membrane (Danielli) showing the general pattern.

Diagram C is a model proposed by Bungenberg de Jong in 1936, in which there is bilayer held together by *electrostatic forces*, the hydrophobic lipid domains facing outward. The diagram represents the particular phospholipid called "lecithin," in which the polar domain contains both a negative charge (filled circle) and a positive charge (open circle), separated by $-CH_2-CH_2-$.

How about the unspecified thickness in the model? The original Danielli-Davson publication in 1935 made no mention of Gorter and Grendel at all. In 1943 the same authors published a book, *The Permeability of Natural Membranes*, which included a chapter on membrane structure, and here the Gorter-Grendel model is shown as well as their own (Fig. 26), but with not a word about theoretical justification. Danielli and Davson appear not to have read or understood Gorter and Grendel's eloquent theoretical arguments.

6. Autopsy. Why Did It Die?

The Gorter-Grendel model died almost as soon as it was born and remained buried for over forty years. Why did this happen? Was it murder, suicide, or accidental death?

Evert Gorter himself must assume some of the blame. He became infected by the excitement of "dynamic biochemistry." This does not mean that he himself turned to biochemical research, but he became involved indirectly, through protein chemistry. Protein molecules were known to be the catalysts ("enzymes") for biochemical reactions, but they were also known to be macromolecules of then almost unimaginable size. If only the structure of these complex molecules were understood, then the secret of life at the chemical level would be revealed. This prospect fascinated many people and drew them into protein research, including evidently Evert Gorter. Because of their size, proteins were almost impossible to study by the conventional methods of organic chemistry, and so little was known about them that *any clue* to protein structure and behavior might throw new light on the problem. It had been discovered that protein molecules could be spread out, apparently as layers of single molecules, on an aqueous surface. This implied that actual molecular dimensions could be measured, following in principle the same procedure that Langmuir had used for fatty acids and other simpler molecules. It also meant that one could then find out how these dimensions change with pH, temperature, or other variables, which were also known to change biochemical catalytic activity.

A paper, with W. T. Astbury and others, published in 1938, illustrates Gorter's enthusiasm. This is not a straightforward monolayer study, but instead describes a study in which more than one thousand monolayers of a protein were piled on top of one another, laboriously, one layer at a time. The reason for doing this was that Astbury wanted to investigate protein structure by means of X rays, a technique that depends on measuring interference between scattered rays from regularly spaced arrays. This particular 1938 paper is unrelated to X rays per se, but has something much simpler to report. The space taken up by more than one thousand monolayers was thick enough to measure directly with a screw micrometer! Since the number of layers was known, this measurement in effect provides the height of a *single monolayer*, an actual dimension of a single protein molecule. The result itself is not important here, but the authors' final comment speaks to the general atmosphere in which the experiment was conducted:

> Needless to say, the thrill of being able for the first time to measure the thickness of a protein chain by such a means far outweighed the satisfaction derived from the more elegant methods.

Evert Gorter had evidently been gripped by protein mania long before this, for he never pursued his initial work on membrane lipids and seemed not to care about the fate of his brainchild, the lipid bilayer. In 1929 his student, F. Grendel, tried to revive interest in the lipid work with a greatly extended account of his thesis research in the *Biochemische Zeitschrift*, an account which includes a lucid elaboration of the theoretical foundation for a bilayer (Fig. 25). But Gorter himself was by then devoting all his energies to proteins, and went on to publish more than fifty papers and review articles on that subject. (Grendel's name has not appeared in the science literature since 1929. At the time of writing, it has not been possible to find out how his career continued.)

There was, to be sure, a hint of foul play as well, at the hands of colleagues from Gorter's own institution. H. G. Bungenberg de Jong, professor of chemistry at the University of Leiden, was younger than Gorter and, to judge from his published work, had an exaggerated sense of self-importance. His work dealt generally with molecular aggregation and included, in several papers published jointly with a student, K. Winkler, some enthusiastic and sweeping (and totally incorrect) speculations about the organization of cell membranes. (For example, the hydrophobic parts of lipid molecules *face out into the aqueous solution* in his models, as illustrated in Fig. 26.) Winkler and de Jong ignore the work of their colleagues in pediatrics, except for a brief reference in which they dismiss Gorter and Grendel's conclusion with the statement that the surface areas they measured for the whole red blood cells could easily have been in error by a factor of two, i.e., their work was asserted to be unable to distinguish between a lipid monolayer and a bilayer. The theoretical support for the bilayer model, which, as just mentioned, had been lucidly reiterated in Grendel's 1929 paper, was ignored. Bungenberg de Jong's work was published in the mid- and late 1930s. There is no record of any rebuttal.

Mostly it was probably neither suicide nor murder, but an accident, the result of connections missed by chance, against a background of unwillingness to think if thinking could be avoided. There was, for example, a mischievous case of the Farish Syndrome which recurred many times in slightly different guises.

Langmuir proved that lipid molecules migrate to the surface of an aqueous solution to form a film there, one molecule thick. The contents of the living cell are an aqueous solution, and the cell manufactures lipid. *Therefore*, the lipid will form a unimolecular film at the cell surface.

That unimolecular film, the argument continues, is most likely the actual fabric of the cell membrane. It seems straightforward and uncomplicated if one has never tried to understand the *theoretical basis* for monolayer formation in Langmuir's experiments.

An American physiologist, H. Fricke, is an example of one who was afflicted in this way. He had published an elegant paper in 1925, the same year as Gorter and Grendel's first paper, in which he used electrical impedance to measure the thickness of the *electrical insulation* of a blood cell membrane. His result, which we now know to have been essentially correct, is 33×10^{-8} cm. Fricke, however, was convinced that the bounding membrane to which this measurement applies must be a single layer of lipid molecules. Grendel, in his 1929 paper, not only reiterated the theoretical impossibility of a monolayer, but also points out that Fricke's thickness of 33×10^{-8} cm would be incompatible with a unimolecular layer on the basis of Irving Langmuir's measured molecular lengths (table 5)—no single lipid molecule could be so long. The criticism provoked no response, from Fricke or anyone else.

Those who casually accepted (or did not object to) the Danielli-Davson model may have been victims of a more modern disease, an adjunct of the explosive growth in scientific progress. The volume of publications has increased to unmanageable levels, and this has engendered a tendency for new ideas to spread by word of mouth. "Have you heard that so-and-so has a really good model for this or that?" It is astonishing how effective *repetition* of a statement like that can be. Lewis Carroll describes the process in *The Hunting of the Snark*:

> Just the place for a Snark! I have said it twice:
> That alone should encourage the crew.
> Just the place for a Snark! I have said it thrice;
> What I tell you three times is true.

And so it is, for any busy listener, without time for critical reflection.

James Danielli had the perfect character to benefit from this. He was the principal author of the Danielli-Davson model and acted as its more or less official spokesman. When he talked (or in his writing) the logic was not always entirely clear, but he made up in personality what he may have lacked in scientific rigor. Danielli was born in London in 1911, the grandson of an artist in stained glass. He was married to a poetess. He had good looks, an abundance of Italian charm, and an infectious exuberance. He was a "man of ideas," in politics as well as science. He somehow inspired trust when he spoke, and people listened.

7. Resurrection

Mercifully, around 1970, the unsupportable models cited above, and others that have not been mentioned, suddenly disappeared. The phospholipid bilayer became recognized as the universal structural framework for all biological membranes, and it is difficult now to understand why it took so long.

R. N. Robertson, in a book called *The Lively Membranes* (1983), describes a meeting held in Australia in the late 1930s. A professor of biochemistry told a physical chemist that he had read about the possibility that cell membranes might be only two molecules thick, and he asked the physical chemist, "Could that be so?" The physical chemist replied that he would be surprised if the membranes were *more than two molecules thick*. (Whose was that voice of wisdom?) According to Robertson, this reply was greeted with "good-humoured disbelief." Robertson goes on to say that "Understanding should have progressed beyond that point, even in remote Australia, because that was approximately 10 years after the classic experiment of Gorter and Grendel."

The historical evidence is that acceptance of the lipid bilayer membrane did not come from rediscovery or reevaluation of the Gorter and Grendel paper, but rather from results obtained by use of more sophisticated (and more expensive) modern tools of investigation. X ray structure analysis and powerful electron microscopes were particularly important in demonstrating that the phospholipid bilayer, with proteins inserted into it, is the all-but-universal framework for cell membrane organization. The Gorter-Grendel paper was rediscovered only after everybody had already been convinced by these other means. The paper is indeed now regarded as a classic—this book is not unique in calling it that—and is often cited as the "foundation" of all membrane science. That, however, is a latter-day resurrection, intended as a teaching device and not as a historical statement. The Gorter-Grendel paper, so obviously well founded upon fundamental physical and chemical laws, is a superb *logical foundation*, even though it failed to be a foundation in the chronological sense.*

*It should be stated, for the sake of completeness, that the natural lipid bilayer is chemically not as homogeneous as shown in figures 25 and 26. It contains not just a single molecule of the phospholipid type, but a heterogeneous mixture of such molecules, differing somewhat in hydrocarbon chain length, and not all containing the same polar "head." How this variable factor in lipid composition may create local variability (e.g., in resistance to temporary deformation) is still an active area of research. In addition, cholesterol is a normal membrane constituent in animal cells, and all functioning membranes have proteins running through them. None of these factors influences the overall molecular organization. A lipid bilayer is always the basic framework.

CHAPTER NINETEEN

Epilogue—The Biological Frontier

WE are now all ardent converts. The membrane is accepted, even honored—given top priority for current research. Anatomists, anesthesiologists, biochemists, cell biologists, immunologists, neuroscientists, opthalmologists, pharmacologists, physiologists—we have all sorts of titles and degrees, but we all now love cell membranes and want to know how they work. We all agree that the phospholipid bilayer, exactly two molecules thick, is everywhere the basic framework of membrane structure, and we understand the forces that tenaciously hold the lipid molecules together. The words "hydrophobic" and "hydrophilic" are ever present, in practically every paper at the molecular level in any of the fields I have here enumerated.

We also now accept that a cell membrane must be more than just a phospholipid bilayer. A lipid bilayer can surround a cell, define it, and segregate its contents from the environment. But, as Overton already told us nearly one hundred years ago, more is needed before we can call a cell "alive." The membrane must be a *mosaic*, containing functional patches provided by protein molecules. The proteins don't line the outside of the membrane, of course, as in the discarded Danielli-Davson model, but they are embedded in the bilayer, often traversing all the way through the membrane, from one side to the other, inseparable from the bilayer unless the whole membrane is disrupted.

The presence of these proteins does not change the basic chemical principles of the assembly of membranes. Protein molecules also have hydrophobic and hydrophilic parts, though their mutual arrangement is far more complex than in simple phospholipid molecules. When incorporated into membranes, the hydrophobic parts of the protein hold the molecules in place, in the hydrocarbon part of the bilayer, away from water; the hydrophilic parts carry out the function, the interplay with the total chemistry of a cell. Figure 27 provides an example, a taste of what may be involved. To go further than this (in terms of structural organization) would take us far beyond the scope of this book.

What kinds of functions do these membrane-bound proteins perform? Taking in fuel and discarding wastes are minimal needs, perhaps adequate for the most primitive unicellular living things. But for the higher multicellular forms of life we need much more.

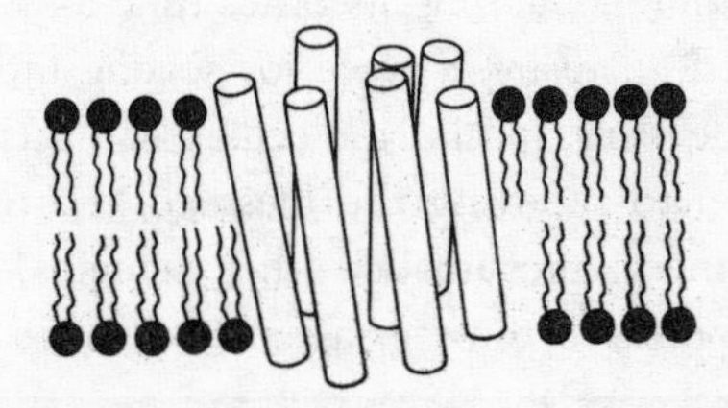

Figure 27. A *hypothetical* protein molecule running through a bilayer in a biological membrane. The bilayer is shown as a two-dimensional cross section. The trans-membrane part of the protein molecule is shown as a bundle of cylinders ("helices"), all of which should be imagined as part of the same molecule, joined together outside the membrane by other segments of the protein.

The surfaces of the bundled cylinders that face out are expected to be hydrophobic, anchoring the overall structure within the bilayer. The inner surfaces of the cylinders are expected to be hydrophilic, capable of transferring ions or polar molecules across the membrane. The cavity formed by the inner surfaces is expected to be "normally closed" except at the instant when actual trans-membrane movement is occurring in response to an appropriate signal.

Where huge numbers of cells act in concert to produce sensory recognition—sight, sound, smell, taste—and to command motion and speech and thought and faith and love, we need rapid communication between cells, sophisticated signals that can pass from one cell to another and to and from the outside world.

And the signals must go through the membranes that bound the cells, because there is no other way in or out!

In other words, the membrane must be a real frontier, where the traffic stops, where one needs to have the right credentials to enter and pass through, or to be allowed to leave messages for the living world inside. There will be orders to stoke up the fires of the metabolic engine or to slow it down, or to start synthesizing something urgently needed elsewhere in the body, and all these orders must come from outside the cell, across the membrane. The membrane, among its other functions, must therefore be what is called a *receptor*, a place for sorting molecules, a place to recognize the very special molecule that acts as messenger. And, in addition to receiving the message, the membrane must have the machinery (presumably other proteins) to execute the appropriate response to the message, and to send it to the inside of the cell.

Sometimes the messenger may not even be a molecule. In the retina of the eye, for example, there is a protein in the membrane (rhodopsin) that can recognize the arrival of a quantum of light. In the ear, there must be a mechanism for recognizing sound waves. In nerve conduction, the first signal is often electrical, and there are proteins with "voltage-dependent" activity.

Research on these matters is still in its infancy. We are finally agreed on *what a membrane is* and can now turn our full attention to *what a membrane does*. But the process has only just begun. The cell membrane is not only a physical boundary, between life and nonlife, but also a temporal boundary, the threshold to a new era of discovery.

Warren Weaver, in *The Story of the Rockefeller Foundation*, writes:

> The century of biology upon which we are now well embarked is no matter of trivialities. It is a movement of really heroic dimensions, one of the great episodes in man's intellectual history.

This was written in 1952, in the midst of the golden age of "dynamic biochemistry," an age which we might label in the present context as the "first revolution" of twentieth-century molecular biology.

One year after Weaver's statement, the structure of DNA was worked out and the molecular basis of genetics became understood. We might call this the "second revolution," the revolution we are now in the midst of, where we shall soon be able to "read" every word of the instructions contained in the chromosomal memory tapes.

For humanity the next advance in biology, the "third revolution," the membrane revolution, may dwarf the other two in its impact. For the dynamic pathways of biochemical synthesis and energy production are nearly universal, apply with only minor differences to all forms of life. The same is true for the chemistry of DNA and the genetic machinery, which is why our emerging gene-splicing industry can use the bacterium *Escherichia coli* as a vehicle for the synthesis of human proteins. In other words, the first two stages of Weaver's great movement discovered facts of life *in general*, without distinction between the humblest orders of bacteria and the highest orders of animals, whereas the stage of molecular biology that lies ahead is ultimately aimed at a higher function, *integrative physiology*, the coordination of billions of cells into a single integrated being. This is the differentiating function that distinguishes not only beast from bug, but also, in its fine detail, man from beast.

We cannot predict exactly where the new era of discovery will lead. Even the things that mankind has probably yearned to understand more than any other—the mystery of memory storage and retrieval and the mechanism of higher intelligence—are within our reach. Like all living matter, the brain consists of cells, and signals across cell membranes will be the operational unit of all of the brain's machinery.

One thing is certain. Our present understanding of what a membrane is, and why it is what it is, is essential for future progress—no good scientist wants to build on a foundation of shifting sand. In my own teaching, laying the groundwork for the next generation of biochemists or biophysicists, I often begin with Benjamin Franklin's experiment on the pond at Clapham Common. Its simplicity and its historical context make it memorable, help the student to understand. I hope this book may do the same beyond the classroom.

One is led to wonder about Ben Franklin himself. At the time of his death—curious as ever about engines and people and mechanics—Franklin had a glass machine in his study, a model to demonstrate the circulation of the blood in the arteries and veins of the human body. Would he have understood that the red color comes from a substance trapped in little cells suspended in the blood? Would he have appreciated the simple relation between the boundary of that cell and the spreading oil on Clapham pond? I like to think that he would have understood it easily, and would have enjoyed the unity of science that could make so marvelous a connection.

Bibliography

General References

Dictionary of Scientific Biography. 16 vol. New York: Charles Scribner's Sons, 1970–80. More than five thousand well-written, scholarly biographies. Each contributor tries to fit his subject into the historical context of his time. This is a wonderful series for browsing—perhaps a better way to learn some history of science than from a more structured book.

Leicester, H. M., and H. S. Klickstein. *A Source Book in Chemistry 1400–1900*. New York: McGraw-Hill, 1952. The history of chemistry is a bewildering subject. This book provides annotated excerpts from original sources, and they help to explain why the subject is so bewildering.

March, Robert H. *Physics for Poets*. 2d ed. New York: McGraw-Hill, 1978. Reprint. Chicago: Contemporary Books, 1983. This is a valiant attempt to educate the general reader. More than half the book is devoted to modern physics (Einstein and later), but earlier physics is by no means neglected.

Singer, Charles. *A Short History of Scientific Ideas to 1900*. Oxford: Oxford University Press, 1959. An excellent concise overview, aimed at the general reader. There is no subject index and there are no original references, but virtually all important scientists are mentioned by name, so that one can go to the *Dictionary of Scientific Biography* for references.

Thackray, Arnold. *Atoms and Powers*. Cambridge, Mass.: Harvard University Press, 1970. A detailed history of the early development of scientific chemistry, from Newton to Dalton. This is "professional" history of science, and may not make chemistry less bewildering for the general reader.

The Earliest Days

Tabor, D. "Babylonian Lecanomancy: An Ancient Text on the Spreading of Oil on Water." *Journal of Colloid and Interface Science* 75 (1980): 240–45. A cuneiform tablet dating from the eighteenth century B.C. describes the use of the spreading of oil on water as a device for prophecy. This is said to be the earliest known written reference to the oil-spreading phenomenon.

Benjamin Franklin

Aaron, Daniel. "Homage to Benjamin Franklin." *Bulletin of the American Academy of Arts and Sciences* 39, no. 4 (1986): 28–46. The author describes this brief essay as the "testimony of a late convert."

Bowen, Catherine Drinker. *The Most Dangerous Man in America (Scenes from the Life of Benjamin Franklin)*. Boston: Atlantic Monthly Press, 1974. This book is limited in scope, but written with great style and enthusiasm. Focuses on Franklin's activities as envoy and mediator, from the Albany Congress of 1754 to his appearance before the Privy Council in 1774.

Carey, M. C. "Benjamin Franklin, Lord Rayleigh, Agnes Pockels, and the origins of surface chemistry." *Falk Symposium* **42** (1984): 5–26. A noted gastro-enterologist sets out on the historical trail. The hydrophobic/hydrophilic dichotomy is a crucial factor in the transport of bile acids into the gut.

Cohen, I. Bernard, ed. *Benjamin Franklin's Experiments and Observations on Electricity*. Cambridge, Mass.: Harvard University Press, 1941. Cohen is the dean of science historians in the United States and tends to write for the professional science historian rather than the general reader. Here he contributes a 160-page scholarly introduction, but the main body of the book consists of a reprint of Franklin's complete papers on electricity.

Doren, Carl van. *Benjamin Franklin*. New York: Viking Press, 1938. Considered to be the definitive biography.

Giles, C. H. "Franklin's Teaspoonful of Oil." *Chemistry and Industry* (1969): 1616–24. This is the solitary commentary on Franklin's paper in the published literature, with photographs of eighteenth-century teaspoons and such. The author is a surface chemist at Strathclyde University (Glasgow), but this account is intended for the general reader.

Goodman, Nathan G. ed. *The Ingenious Dr. Franklin (Selected Scientific Letters)*. Philadelphia: University of Pennsylvania Press, 1931. A grand selection, demonstrating the breadth of Franklin's interests in all kinds of practical and theoretical questions.

"Of the stilling of Waves by means of oil." *Philosophical Transactions of the Royal Society* 64 (1774): 445–60. The complete text. About half of it is quoted in this book. An advantage of looking at the original is that it allows browsing among other letters published the same year.

Stearns, Raymond P. "Colonial Fellows of the Royal Society of London, 1661–1788." *William and Mary Quarterly*, 3d ser., 3 (1946): 208–68. The word "colonial" here refers to all the American colonies, including West Indian islands.

Joseph Priestley

Holt, Anne. *A Life of Joseph Priestley*. Oxford: Oxford University Press, 1931. Priestley the stubborn nonconformist. The focus is on the man and his friends and enemies.

Schofield, Robert E. ed. *A Scientific Autobiography of Joseph Priestley*. Cambridge, Mass.: MIT Press, 1966. Excerpts from Priestley's voluminous scientific correspondence.

The Eighteenth Century

Boswell, James. *The Private Papers of James Boswell.* Edited by Frederick A. Pottle et al. New York: McGraw-Hill, 1950–77. The first volume is *London Journal, 1762–1763*. A later volume describes the tour to the Hebrides with Samuel Johnson in 1773.

Hankins, Thomas L. *Science and the Enlightenment*. London: Cambridge University Press, 1985. A clear and concise account that should not be too difficult for the general reader.

Newton, Isaac. *Opticks* 4th ed. (1730). Reprint, with preface by I. Bernard Cohen. New York: Dover Publications, 1979. This is the most readable of Newton's works. The "queries" at the end of the book are Newton's way of saying what he thinks but can't prove.

Schofield, Robert E. *Mechanism and Materialism*. Princeton, N.J.: Princeton University Press, 1970. A detailed history of British science in the eighteenth century.

Pliny the Elder and Younger

Greig, Clarence. *Pliny: A Selection of his Letters*. London: Cambridge University Press, 1978. Delightful translations.

Rackham, H. *Pliny: Natural History*. 10 vols. Cambridge, Mass.: Harvard University Press, 1938. This is one of several translations available in most libraries. It has the Latin text and its translation on facing pages.

Lord Rayleigh and Science in His Time

Aris, R., H. T. Davis, and R. H, Stuewer, eds. *Springs of Scientific Creativity*, Minneapolis: University of Minnesota Press, 1983. This collection of essays includes an enthusiastic biography of Lord Rayleigh by John N. Howard and an excellent chapter on Maxwell by C. W. F. Everitt. The remark of H. A. Rowland about Maxwell that is quoted in chapter 12 is taken from this source.

Bragg, W. L. and G. Porter, eds. *The Royal Institution Library of Science: Physical Sciences*. 10 vols. Barking, Eng.: Elsevier, 1970. This is a selection of the Friday evening discourses held at the Royal Institution in London between 1851 and 1939. The Friday lectures were intended to be comprehensible to an audience without expert scientific knowledge. Tradition called for "talks distinguished by simplicity and clarity and, when appropriate, illustrated by interesting experiments and demonstrations." The ten volumes are a treasure trove of discovery, including many lectures by Faraday, Tyndall, Rayleigh, J. J. Thomson, and so on. J. J. Thomson announced his epoch-making discovery of the electron in one of these discourses in 1897.

Kurti, N. "Helmholtz's Choice." *Nature* 314 (1985): 499. This note briefly recounts the offer of the Cavendish Professorship to Helmholtz in 1871. An earlier note by the same author, "Opportunity lost in 1865?" *Nature* 308 (1984): 313–14, discusses Helmholtz's candidacy for a similar position at Oxford University in 1865. The man chosen to fill that position (R. B. Clifton) did not fulfil expectations, but held the position for fifty-three years. This accounts for the supremacy of Cambridge over Oxford in the history of physics for the period covered in chapter 9.

Sharlin, H. I. *Lord Kelvin*. University Park: Pennsylvania State University Press, 1979. An interesting biography, with a serious attempt to portray the influence of dramatic changes in ideas on the man who was the most dynamic of the physicists of his time.

Strutt, John William. *Scientific Papers by Lord Rayleigh*. 6 vols. Reprint. New York: Dover Publications, 1964.

Strutt, R. J. *Life of John William Strutt, Third Baron Rayleigh*. Madison: University of Wisconsin Press, 1968. A low-key biography by Rayleigh's son (the 4th baron).

U.S. Hydrographic Office. Bureau of Navigation. *The Use of Oil to Lessen the Dangerous Effect of Heavy Seas*. Reports 82 and 83. Washington, D.C.: GPO, 1886 and 1887.

The Hydrophobic Effect

Tanford, C. *The Hydrophobic Effect*. 2d ed. New York: John Wiley and Sons, 1980. This book is addressed to the professional, but the book is quite short, and some chapters are lighter than others.

Agnes Pockels

Giles, C. H., and S. D. Forrester. "The Origins of the Surface Film Balance." *Chemistry and Industry* (1971): 43–. This article deals mostly with Miss Pockels and includes considerable research by the authors on her life and family.

Ostwald, W. "Die Arbeiten von Agnes Pockels uber Grenzschichten und Filme." *Kolloid Zeitschrift* 58 (1932): 1–8.

Irving Langmuir and Surface Science

Adam, N. K. *The Physics and Chemistry of Surfaces*. 3d ed. Oxford: Oxford University Press, 1941. A professional book; a scientific education is necessary for more than superficial reading.

Langmuir, Irving. "The Constitution and Fundamental Properties of Solids and Liquids. II. Liquids." This is the awkward title of Langmuir's paper on molecular orientation in *Journal of the American Chemical Society* 39 (1917): 1848–1906. The paper is of course included in the *Collected Works*, but it may be more readily accessible in the original journal.

Langmuir, I. "Pilgrim Trust Lecture. Molecular layers." *Proceedings of the Royal Society* (1939) **A170**, 1–39. One of the lectures in which Langmuir correctly explains the hydrophobic principle, but misguidedly supports the protein model advocated by Dorothy Wrinch.

Suits, C. G., *The Collected Works of Irving Langmuir*. 12 vols. New York: Pergamon Press, 1960–62. Volume 12 contains a biography by A. Rosenfeld, "The Quintessence of Irving Langmuir," which is also available as a separate book. Other tributes to specific aspects of Langmuir's work by his scientific colleagues are scattered through most of the twelve volumes.

Biological Science

Baldwin, Ernest. *Dynamic Aspects of Biochemistry*. London: Cambridge University Press, four editions from 1947 to 1963. The golden age of biochemistry, captured in mid-life.

Bayliss, William M. *Principles of General Physiology*. London: Longmans, Green, four editions 1914–24. A book of unusual breadth, written with gusto and a fine sense of what is important. A good introduction to functional biology, despite the fact that some of Bayliss's ideas have proved to be wrong. Bayliss died in 1924, and posthumous editions of his book, revised to more conventional "textbook" style, are not the equal of the original.

Fruton, Joseph S. *Molecules and Life*. New York: Wiley-Interscience, 1972. A labor of love, with perhaps a little too much loving detail.

Henderson, Lawrence J. *The Fitness of the Environment*. New York: Macmillan, 1913. Reprint. Gloucester, Mass.: Peter Smith, 1970.

Molecules of Life. Scientific American 253, no. 4 (1985): Special Issue. Colorful pictures of models of DNA, RNA, and protein molecules, and fanciful representations of membranes, cellular skeletons, and the like.

Robertson, R. N. *The Lively Membranes*, London: Cambridge University Press, 1983. From Langmuir to the present day. An eminently successful effort to reach the general reader, with many references for those who want specialized detail.

Ernest Overton and Evert Gorter

Collander, P. R. "Ernest Overton (1865–1933): A Pioneer to Remember." *Leopoldina*, 3d ser. 8 and 9: 242–54. A brief but affectionate biography.

Gorter, E., and F. Grendel. "On Bimolecular Layers of Lipoids on the Chromocytes of the Blood." *Journal of Experimental Medicine* **41** (1925): 439–43.

Overton, E. *Studien über die Narkose*. Jena: Gustav Fischer, 1901. This is the book for which Overton is remembered. An English translation by Robert L. Lipnick is forthcoming.

Zwaal R. F. A., et al. "The Lipid Bilayer Concept of Cell Membranes." *Trends in Biochemical Science* 1 (1976): 112–14. A brief retrospective critique of the Gorter and Grendel experiments.

Index

Popular Science from Oxford

Nature's Building Blocks: An A-Z Guide to the Elements
John Emsley

A readable, informative, fascinating entry on each of the chemical elements, arranged alphabetically from actinium to zirconium. A wonderful 'dipping into' source for the family reference shelf and student.

'What for many might be a dry and dusty collection of facts has been turned into an amusing and finely crafted set of mini-biographies . . . . This is a fine, amusing and quirky book that will sit as comfortably on an academic's bookshelf as beside the loo. . . .'

Nature

Molecules at an Exhibition: Portraits of Intriguing Materials in Everyday Life
John Emsley

What is it in chocolate that makes us feel good when we eat it? What's the molecule that turns men on? What's the secret of Coca-Cola? This fascinating book takes us on a guided tour through a rogues' gallery of molecules, from caffeine to teflon, nicotine to zinc.

'A fine example of popular science writing at its best. It is educational, interesting, may prove inspirational and therefore deserves to find a very wide readership.'

Times Higher Education Supplement

Popular Science from Oxford

The Last Word
New Scientist

Why is the sky blue? Does it really get warmer when it snows? Why doesn't superglue stick to the inside of the tube? How is it possible to uncork a bottle by hitting the bottom? Can you drive through a rainbow? Should you walk or run in the rain? Why does soap make bubble bath bubbles collapse? Fun and informative, this is entertaining reading for anyone who has ever puzzled over these mysteries of life.

The Last Word 2: More Questions and Answers on Everyday Science
New Scientist

Why do boomerangs come back? Why does grilled cheese go stringy? What would happen to a pint of beer in space? More questions and answers from *New Scientist*'s popular column make a fun, fascinating, and enlightening read for all.

OXFORD